NATURE'S WINDOW

Nature's Window

D. Bruce Swinehart, Jr.

American River College
Sacramento, California

Drawings by
Jo Smith

KENDALL/HUNT PUBLISHING COMPANY
Dubuque, Iowa

Cover photo by D. Bruce Swinehart, Jr.

Library of Congress Catalog Card Number: 76-18672

ISBN 0—8403—1532—5

Printed in the United States of America

Too often we find ourselves so involved with the mere mechanics of living that we lose sight of what living can offer. If one can just find the true beauty that nature offers, life will be one thrill after another.

CONTENTS

Acknowledgments

The inspiration and encouragement for this book are the result of the influence of many people. Friends, family and colleagues provided references, philosophy, information or inspiration whenever needed. Mr. Peter Hayes of *The Sacramento Union* must be credited with the idea of the Nature's Window column that appeared for nearly four years in that paper. Many of the articles in this book appeared in *The Sacramento Union.* Mr. Chester Harris, the late Bruce Nicols, and several other special editors provided fine cooperation in all of those years.

The environmental ideals expressed in this book were greatly influenced by the National Audubon Society and particularly the Sacramento Chapter.

The important part students have played in this manuscript must be mentioned. They unknowingly often provided interesting subject ideas.

Jo Smith, the illustrator, gave far more than her talent of illustration. Her creativity was evident and her easy-going manner was helpful.

To all the people listed above I wish to express my indebtedness and sincere appreciation. To my wife, Jeanne, for her unvarying criticisms, untiring effort, and constant support I express my gratitude.

About the Book

This book is a compilation of most of the columns called Nature's Window that appeared between Nov. 1970 and Nov. 1974 in *The Sacramento Union* newspaper.

The potpourri of nature subjects reflect the attitudes and philosophy of the author and present interesting information in a manner consistent with the author's personality.

The articles are written in a style that makes them easy to understand and appreciate. The subject matter is most appropriate for those familiar with or living in the central valley of California. There is, however, much general information that is of interest in almost any locale in or out of California.

The subjects within each chapter are arranged in an order that will be useful rather than taxonomically accurate.

It is hoped that this book will add to the knowledge of those who already have an interest in nature and will interest those who are looking through Nature's Window for the first time.

1
Seasons

In Spring the fresh green buds appear to indicate a reawakening of the life-giving processes of nature just as the leaves falling from the trees indicate that nature needs a rest from the cyclic chores of harnessing the sun's energy.

April Showers

The words from the lovely song "April Showers" provide us with a truly ecological promise when the melodious strain states: "Though April showers may come your way, They bring the flowers that bloom in May."

For one who is interested in ecology the next questions are, why do the showers arrive in April and do they really bring the flowers? Understanding climate and its variations is a very complex subject, especially in our valley. I do not have to point out that our showers do not start in April but begin in the fall and continue sporadically through spring. April showers for us are just a part of our wet season.

Our annual cycle of seasons is similar to the rest of the Northern Hemisphere, but the cycle is altered by various land features. Usually a low pressure area exists over the Northern Pacific Ocean that contributes to our winter weather, and a high pressure area moves in that helps cause our summer weather. The reason for the shift involves the tilt of the earth on its axis and the alternate cooling and warming of Arctic lands. This causes the rise of warm moist air over the ocean in winter. Prevailing winds set in motion by the earth's rotation move in a west to east direction over California. Some of the rain is stopped by the Coast Ranges in winter and almost all is kept from the valley in summer. The dry summers and mild but wet winters cause many to call our climate Mediterranean.

Anyone who looks through nature's window soon learns that climate determines the type of plants and animals that are found in an area. If one thinks again of April showers, he may discover that it is the length of the day and the warmth of the sun that plays the largest part in bringing May flowers.

Tule Fog

There are many virtues for those of us that live in the Central Valley.

Our close proximity to the mountains and the ocean are but two, while the mildness of climate is another.

The blazing sun of summer and the tule fog of December and January are not generally considered virtues however. The many days and nights when tule fog blocks the warming rays of the sun from our valley is, perhaps, the most difficult aspect of our valley climate to accept.

The term tule is used because the fog is often associated with swampy areas where tules grow. The correct term, however, is radiation fog, and it occurs when weather conditions are just right.

When the earth radiates heat at night the moist air over the land may condense into fog.

Sometimes cold air from the surrounding hills or mountains drains into valleys causing the moisture-laden air to condense into fog. If there is a slight breeze, the fog created may be extensive.

If there is no wind, the fog may be a thing of beauty extending from a few inches to several feet above the ground. If the morning sun is bright and the sky clear, the thin bands of tule fog can be a fascinating sight.

This type of effect is often used in movies and television and can have intriguing effects on the imagination.

As a child, I remember riding my bicycle through gossamer bands of fog that completely concealed the bicycle and left my head in a world of its own.

I can remember the exhilarating feeling to this day. Even fog can be a thing of beauty if one has the proper perspective.

Winter Preparations

When the first snows of winter have fallen in the high mountain areas of our state, it is a positive sign that the full force of winter will soon be upon us. That means it is time to check our furnace filters, put antifreeze in our cars, shop for winter clothes, along with many other mundane chores.

Nature also prepares herself for the harshness of winter, often in some very bold ways.

A trip to the mountains will invariably lead to the admiration of the beautiful leaves that are ready to fall from the oaks, dogwoods, maples, cottonwoods and aspens. Their work for the tree finished, they shed their green work clothes and the vibrant colors of fall are unmasked. Since they are no longer needed, the softest breeze causes them to flutter to the earth and await for the infinitesimal life-renewing processes. With the leaves of the tree gone and resistance to the wind lessened, the food-making process is stopped. The food has been stored for the growth of spring leaves. The tree is now ready for the extremes of winter.

The raccoon prepares himself for winter by eating all he can. Consequently, he developes huge layers of fat. He is so fat he may actually waddle. When food becomes scarce, he will have to exist upon his stored fat and the little he can scrounge. He will sleep through very rough spells of weather, but he will not hibernate.

Squirrels, too, have developed some layers of fat, but they have also cached much of the abundant food supply for later use. When the terrain is blanketed by the durable snows of winter, they go into the deep, energy-conserving sleep called hibernation. Their body temperature lowers considerably, and their heartbeat slows greatly. In this state they normally do not awake unless they are warmed. Little energy is used in this non-productive time of year.

Some other animals, like deer and many birds, leave the mountains and north for areas that are not so cold and inhospitable. Our valley provides its visitors the much needed relief from these barren winter lands.

Autumn Leaves

The splendor of nature's finery is especially obvious in fall.

The lush green of spring and the cheering warmth of summer are followed by a feast of colors.

I hope everyone appreciates the life- sustaining abilities of the green-colored chlorophyll that dominates the leaf's hue for almost all of the life of the leaf.

Chlorophyll, however, is but one of thousands of pigments found in plants. Many of the others undoubtedly serve to attract insects and other animals for the purpose of pollination. Chlorophyll, for instance, is the only pigment found in many wind-pollinated flowers.

Carotenoid pigments number more than 60 in various kinds. They are broken down into two groups, the carotenes and xanthophylls, which range from lemon yellow to tomato red.

The carotenoids are so durable as compounds they may survive the digestive processes of animals and wind up as the color in egg yolk and butterfat.

The third group of pigments found in plants is the anthocyanins. They produce colors from the palest pink through red to blazing purple. These pigments are found in solution in the plant's individual cell material and no vital function has been attributed to them. It is known that they are readily affected by acidity and alkalinity.

The leaf color of fall is brought about by a chemical reaction in which pigments and weather play an equal part. As the life of the leaf ends, the production of chlorophyll stops, and the more durable yellow carotenoids and red anthocyanins reveal themselves in their exotic hues.

The sharp drops in temperature make the changes more dramatic.

Leaf Fall

I wonder how many people just accept the fact that leaves fall from a tree and never question why. For most of us the falling of the leaves is the most obvious signal of the changing seasons outside of the weather. In spring the fresh green buds appear to indicate a reawakening of the life-giving processes of nature just as the leaves falling from the trees indicate that nature needs a rest from the cyclic chores of harnessing the sun's energy.

It is easy to attribute to the wind the often colorful accumulation of leaves that move like tiny bouncing clowns in response to a happy enthusiastic audience. It is not simple, however, to explain why the wind is suddenly able to blow the leaves from the tree in the fall but not before. In the spring and summer the wind may blow violently but scarcely a leaf is wrested from the tree.

What is the signal Mother Nature sends to her leafy representatives so they shed their leaves as part of the constant recycling that keeps the living world in working order?

Most botanists believe the trigger for most trees is the shortening of the day. Just how this activates the changes that follow is not completely understood. It is believed that because of the reduction of the life-giving light, the process that only the green plant can perform (photosynthesis) is curtailed, and the growth stimulators (auxins) are reduced. In one of nature's many intricate ways this causes a new layer of tissue to develop at the base of the stem of the leaf. This tissue bears the curious name, abscission layer, which means cut off. Soon after the new layer develops, it starts to deteriorate. This weakens the hold the branch has on the leaf until it is primarily held by the former life-giving water system now rendered useless.

The wind or frost merely administers the *coup de grace* in the preparation of the tree for winter. With its leaves gone, resistance to the rain-soaked winds is lessened, and the chances of survival through the winter greatly improved. As in all things seen through nature's window there is a purpose and a plan.

2
Fish, Amphibians, Turtles, Lizards and Snakes

Many who casually look through nature's window attribute "good guy" and "bad guy" roles to various of nature's players.

As one learns more about nature, he finds it is only through man's eyes that good and bad judgments are made.

Carp

Our valley in summer encourages many people to spend their leisure time in or near the water. Swimming, water skiing, rafting, canoeing, and fishing are but a few of the many activities where we often see many small fish. It is common practice to call these small fish minnows, which may or may not be correct. The term minnow can only be applied to fish of the minnow family or Cyprinidae. There is a great variety of Cyprinids, however, and many of the small fish seen are true minnows. Young, free swimming fish can better be called fingerlings.

One of the very common fish that we may see in large numbers in our valley waters is the carp, *Cyprinus carpia.* This member of the Cyprinid or minnow family has the typical minnow characteristic of the pelvic fins being in the abdominal area. The pelvic fins are the hind or lower pair of fins located on the belly of the fish. In the minnows they are located back near the anus.

The carp, itself, has quite large scales and is a grayish, greenish or goldish color, usually darker on the back. There are two little barbels or whiskers near each end of the mouth and a heavy saw-toothed spine at the front of the dorsal fin. The tail or caudal fin is deeply forked.

A young carp may be 6 to 8 inches long in one year and 12 to 15 inches in two years. Weights to 42 pounds have been reached in America and 90 pounds in Switzerland. A six-pound fish may lay two million eggs which it does in shallow water.

Carp were deliberately introduced into California in 1872 when they were thought desirable. Now they are considered a pest since they eat plants that waterfowl eat, roil the water and drive out silt sensitive fish and compete directly with game fish. They are edible but bony.

A recipe often quoted for carp is to split the fish open and pin to a plank. Rub with butter, lemon, salt and pepper. Then broil in front of an open fire. When golden brown, take the fish off the plank. Throw away the fish and eat the plank. I guess that tells you something.

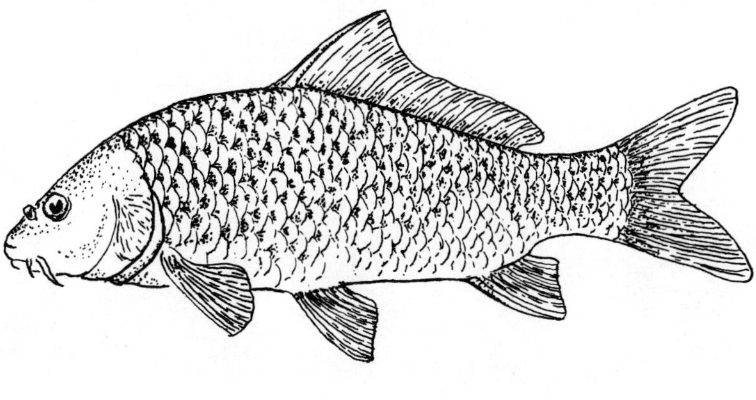

Salmon

Fish watching can be very entertaining and easy to do in our area. A visit to Nimbus Fish Hatchery in the fall will enable you to see the large salmon returning from the ocean to spawn and die. The sight of these large fish jumping to get up the ladder is quite impressive. After they have reached the holding tanks at the top, they create an unforgettable scene.

Although there are five kinds of salmon found along the Pacific Coast, the king salmon, *Oncorhynchus tshawytscha,* is by far the dominant one in the Sacramento-San Joaquin River complex. They are rarely taken from the ocean south of San Luis Obisbo County. The Sacramento and San Joaquin Rivers are the southernmost rivers that support large runs.

The king salmon, sometimes known as chinook, quinnat, or spring salmon, enters the streams and rivers during two periods of the year. If they enter during the spring, they spend the summer months in deep holes until fall. Then they move to the spawning grounds. Those that enter in fall swim directly to the spawning grounds.

After the eggs are laid, fertilized, and covered with gravel, the parents soon die. Depending upon the temperature of the water, the eggs generally hatch in 35 to 55 days. Evidently, the young fish migrate down the rivers within a few months after hatching. During this period, large numbers of young are carried into irrigation canals and eaten by predators. Those that reach the ocean disperse rapidly and seem to disappear for three or four years.

Most salmon return to the river to spawn. When entering the river they no longer feed but become very pugnacious. They will strike at lures, other fish, and the like.

Visit Nimbus Fish Hatchery during the fall salmon run. I can assure you the effort will be well worthwhile. It is located east of Rancho Cordova off Highway 50.

Bullfrog

When one looks through nature's window all of the participants are not visible. Some of them are very evident, however, by their calls. An excellent example of this is the bullfrog, *Rana catesbeiana.* Their booming voices saying "jug-o-rum" resound across the water at night in a chorus that is very characteristic of almost all our waterways.

This large amphibian may have a body length of eight inches, and the hind legs are often up to ten inches long. The back is green without warts. The ear located behind the eye is as large or larger than the eye. This extremely aquatic frog seldom is found far from water. They eat insects, earthworms, snails, small fish, frogs, birds and mammals. Eggs are laid in a mass up to a yard in diameter and may contain 10,000 to 20,000 eggs.

The larvae remain as tadpoles, and sometimes reach 5 1/2 inches in their two years. Their food is aquatic plants, oozes and insects. They have irregularly spotted tails and yellow bellies.

The bullfrog is the frog generally jumped in the frog jumping contest at Angels Camp. Most people assume that it is the frog referred to by Mark Twain in his famous story written in 1865, "The Celebrated Jumping Frog of Calaveras County," and perhaps it is. If it is, however, the story could not have taken place at Angels Camp since the bullfrog was not introduced into California till 1905 at the earliest. The natural home of the bullfrog was east of the Rocky Mountains.

Mark Twain was very familiar with the bullfrog from his Mississippi days. Wouldn't it be disappointing if the incident happened somewhere other than Calaveras County? If the contest were real and happened in Angels Camp, our native red-legged frogs, *Rana aurora,* would have to have been the participants. How about that?

Tree Toad

One of the characteristic signs of approaching summer is the haunting sounds of the night creatures. Their melodies are so much a part of our expectations that as summer progresses, we may not be conscious of their presence. Such is not the case in spring when their voices first fill the air. During one warm evening, a tiny tree toad voiced his views to the world right outside my front door. No one could ignore his outcries.

The Pacific tree toad, *Hyla regilla,* or as it is often called, tree frog, is a tiny creature whose color is variable and changeable. Some are green, brown, reddish, tan, grey or black often with dark body spots and a dark *y* or *t* shaped mark on the head. The underparts are usually a whitish or pale yellow without spots. Males have an olivaceous to dusky throat. This frog is best told by a conspicuous blackish eye stripe that extends from the nostril to well behind the eye.

The body of this vocal little fellow is seldom over 2 inches long. He has a sticky disc on the tip of each toe that enables him to climb or cling upside down to a surface such as glass. Although largely nocturnal, the Pacific tree toad is also active in daylight. He catches insects with his darting tongue usually on or near the ground. He has no particular affinity for trees even though his name and ability imply he does.

The most impressive ability of this tiny creature is the volume of his voice. The booming krerk-ek is all out of proportion to his size. He will sing at one-second intervals in most situations when temperatures are about 68 degrees. When one hears a chorus of them, most other sounds are excluded.

They take refuge in rock crevices, under bark, in rodent burrows, on vegetation and other moist protected places from the valley floor to the high Sierra. One can find them from British Columbia to Lower California, usually associated with water.

The voice of this widespread little amphibian is perhaps better known than he is. Many casual viewers through nature's window hear the loud voice and assume it is from the bullfrog. Are you one?

Western Toad

Quite often in the summertime we are surprised by a creature that we know is harmless but nevertheless causes a real startle reaction. It may be when we walk across a freshly sprinkled lawn at night, or when we move rocks or boards in our garden. I am referring, of course, to the western toad, *Bufo boreas.*

This toad is most active at night and usually hides in the daytime. To hide, it burrows in the moist earth usually under a surface object. If living conditions are right, it may not move far and will return to the same hiding place night after night. This species is prevalent throughout California from the valley to high mountains.

All toads are amphibians which means they must have water or moist conditions in which to lay their eggs. Two strings containing up to 16,500 eggs are laid in shallow water no deeper than 12 inches.

The adult western toad is from 2 1/2 to 5 inches in length excluding the legs. The very warty skin, small eardrums, and prominent parotid glands distinguish toads from frogs. The parotid gland is the large swelling or bump directly behind each eye. The general coloration is grayish, blackish, brown, or dull green above with a white stripe down the backbone.

The toads are very useful in the garden because they eat many pests such as the cutworm, potato beetles, chinch bugs, ants, and slugs. Prey is captured by a two-inch sticky tongue.

Despite its homely appearance, it is not harmful to man. It can exude a slime distasteful to many enemies, but does not cause warts or blindness in man or dogs. The Chinese used toad skin for a medicine probably because of the adrenalin present. It cannot live for years confined in a rock as some people think. Even though Shakespeare wrote that there was a jewel in the head of a toad, he did not mean an actual jewel, but that even an ugly creature like a toad had some redeeming qualities. An excellent ecological thought for all of us.

Desert Tortoise

An occurrence that both amazes and disturbs me is the number of desert tortoises, *Gopherus agassizi,* that I have found or heard about in our area.

I am sure some people erroneously believe they are one of our local animals. The tortoises, however, have probably been illegally brought here from their native home in the arid southwestern region of the United States.

In past years, they were also sold in pet stores. Buyers were often led to believe they would control insect pests in their yards.

Desert tortoises are strict vegetarians, however, and eat only plants including the flowers. They also dig and make a nuisance of themselves in general. Their digging often allows them to escape.

Normally, they will not survive in our climate unless they are brought inside and kept well above our minimum temperatures in winter.

Adults reach about 13 inches in their shells or carapace length. The carapace is high and has pronounced growth rings on its shields. These growth rings cannot be used to indicate age.

The underside or plastron has six pairs of shields that may also show growth rings. These are often worn smooth. The head is covered with flat scales, and the limbs are stocky with no sign of webs between the blunt nails. The carapace is dull brown, but the shields often have light brown or yellowish center spots.

When two tortoises meet they bob heads. If they are both males, a battle may ensue and each attempts to turn the other over on its back. An overturned male will soon die unless it finds something to push against to right itself.

The desert tortoise is the largest, by weight, of California's reptiles and has been selected as our state reptile. It is now protected by law, and it is illegal to possess them.

In spite of this, many are killed deliberately by shooting and others are killed accidentally by cars.

The advance of civilization into the desert is also reducing their habitat. Let us hope the state reptile does not follow the grizzly bear, our state mammal, into extinction.

Pacific Pond Turtle

The only true native turtle in our valley is the Pacific pond turtle. Because there are more than 300 species of turtles in the world, it is difficult to understand why so few species are found here, when many species are found in the eastern United States.

The Pacific pond turtle, *Clemmys marmorata,* has a shell or carapace length of up to six inches. The carapace is quite low in outline from top to bottom and somewhat oblong in outline. It has no median ridge and becomes smoother as the turtle grows older.

Younger turtles usually have concentric ridges in the carapace shields which gives them a rougher appearance.

The underside or plastron has six pairs of shields and is yellowish with dark blotches.

The basic color of the carapace is dark brown, blackish, or olive color. Each shield is marked with a network of spots, lines or dashes of brown or black that tend to radiate from growth centers.

The markings sometimes unite to form a distinctive branched network.

The Pacific pond turtle is thoroughly aquatic, although they leave the water to bask in the sun.

The females also leave the water to lay their eggs in an earthen cavity they dig.

The eggs have a hard white shell and are about an inch in size. They are usually laid in a sunny place along rivers, streams, ponds, or even some distance from the water. The eggs are covered with soil.

Food is essentially water plants in some areas but insects, carrion, fish, earthworms, etc., are more important in others.

Pond turtles seem to prefer the quiet water of ponds, small lakes and slow-moving streams, but they are also found in most suitable waters. They have even been recorded in brackish and salt water.

Because turtles were sold in pet stores for years, many unusual species may be found here.

After handling aquatic turtles, your hands should be washed well to prevent various infections. Caution children especially.

Foothill Alligator Lizard

There are a lot of surprises in store for those that poke around in their yards and gardens in the spring. When piles of weeds, rocks, boards or boxes are moved, there is often some creature hiding below. It would not be unusual to find Jerusalem crickets, black widows, sow bugs, centipedes, or even an alligator lizard resting in a winter hiding place.

In our valley the foothill alligator lizard, *Gerrhonotus multicarinatus,* is not at all uncommon. Watchers through nature's window often see this sleek, shiny lizard scurrying through the leaves and grass. Even when they do not know its name they probably would say it looks like an alligator. Its large body scales, triangular head, and long slender tail make identification very easy.

The brownish or olive-colored body has various dark markings, and the large ear opening is quite obvious. The legs are rather short, and when this deliberate-acting fellow is in a hurry he moves his body in a somewhat sinuous snakelike movement. His forked tongue is often used to sample the environment just as snakes do. I have seen 16-inch specimens, and although they appear large, their long slender tail makes up two-thirds of their length.

Generally, the alligator lizard searches for food in the daytime, but on warm nights he is also active. Beetles, grubs, spiders including black widows, small lizards and mammals are typical food.

This lizard, although harmless, is quite ferocious. He will twist and squirm and try to bite when held. If he does bite, he will probably not draw any blood, but the pinch will leave quite a dent. I have actually seen them dent a pencil. I often let them bite me on the thumb with one jaw on the thumbnail. Sometimes they really hang on to the point of embarrassment.

The long tail is an interesting protective feature and is usually kept in motion. It serves to draw the attention of any predators watching. If they attack, the tail is generally grabbed and will break off easily. Once severed, the tail does a long series of convulsive jerks that hold the attention of the predator. The lizard quietly crawls away.

Watch for this interesting reptile and be glad he is helping you control many of your garden pests.

Garter Snake

There are many things in the out-of-doors that scare some people. Most fears are unfounded and have no real basis in fact. Especially is this true in regard to a fear of snakes. Of all the everyday things that can cause us harm like automobiles, diseases, electricity, airplanes, pesticides, poisons, etc., it is very inconsistent to have such strong fears about harmless snakes. The only snake in our valley that is dangerous is the rattlesnake, and all of them have rattles or at least a blunt tail. One of the most common harmless snakes that we have in our valley is the garter snake. There are two kinds we see in our valley—the common garter snake, *Thamnophis sirtalis* and the western garter snake, *Thamnophis elegans.*

Both snakes may get to be 48 inches in length and have a slender body and a tail that tapers to a sharp tip. The upper surface of the body is blackish or grayish, and the typical specimen has a yellow or orangish line down the middle of the back and sides. There may or may not be red in the pattern on the sides.

The common garter snake is generally found near water but eats both land and water animals, and if the rows of scales are counted around the middle of the body, they should number 19. The western garter snake normally has 23 of the scale rows and is found in two forms. One form lives mainly on land and feeds on slugs, salamanders, and mice. It rarely eats fish. The other form is aquatic and feeds on fish, amphibians, and water insects. There is no stripe on the back, but there are light spots instead.

Both of the species can discharge a very foul-smelling substance from a gland in the anus if handled or disturbed. Soap and water will remove the smell easily from your hands. Young snakes are born alive, and there may be more than 20 in a brood.

Garter snakes are often called gardener snakes or water snakes. Both names are somewhat appropriate but not accurate. The next time you see a garter snake count the scale rows and see if you can tell which one it is. You will feel like a real expert.

Gopher Snake

The coming of fall causes many of nature's creatures to change their ways of tapping nature's larder. These changes are often noticed by one who investigates the intricate but often simple pattern of life. Many times, however, the evidence of change is curiously subtle. It is very obvious to anyone looking through nature's window that the shiny, often beautifully decorated butterfly chrysalis is the insect's way of passing time until it can again count on the bountiful smorgasbord of spring.

Every fall I am called upon countless times to identify a tiny creature that often creates panic in the beholder.

I can understand the misgivings of the inexperienced, but it seems so needless when so little knowledge should dispel any fear. This tiny creature writhes, coils, hisses and strikes and does everything in its power to frighten all who threaten it. These, however, are its only weapons.

It can strike but not wound. It can hiss but not wither. It can coil but not crush. All it has, it would seem, is its ferocious and frightening behavior to keep it safe. There is one other defense it has been blessed with or perhaps it is a curse.

This tiny bit of life has a coloration and pattern somewhat like a rattlesnake and although he is quite different, many are killed each year by those who fear they may be poisonous.

I am sure many have already guessed I am referring to the very beneficial gopher or bull snake. In the fall, the small mice, lizards, and insects are hard to find, and the young snakes venture out into areas they normally do not frequent. Here they are seen and often killed.

It is vital to realize that every rattlesnake has either a rattle or button at the end of its tail regardless of age, and that all other snakes in this area are harmless. The gopher snake, large or small, is a part of nature's scheme and should be appreciated by all.

Common Kingsnake

Though many people have fearful feelings about snakes, most recognize the kingsnake as being beneficial. The kingsnake is well-known for his cannibalistic tendencies to eat other snakes, even rattlesnakes. He kills these snakes by constriction and stops their breathing. He does not crush them or deliberately break their bones. This is done, as it is in other snakes, during the swallowing process.

The kingsnake we find here in our valley is the common kingsnake, *Lampropeltis getulus.* In past years it was sometimes called boyle or chain kingsnake. This species ranges from Florida to southern New England and with several other species extends west to Nevada, California, lower California, and into Mexico. It is found up to the middle elevations in our mountains.

Our typical adult specimens are two and one-half to three and one-half feet long. Some have been found over five feet. The typical color pattern is a series of rings that alternate cream or white with black or maroon.

There are many variations to this pattern that may be localized in certain areas. The variations may run from longitudinal stripes down the back to a speckled all black.

This rather quiet reptile eats not only other snakes but consumes lizards, birds, bird eggs, mice, and pocket gophers. It is even known to climb trees after prey. It is immune to rattlesnake venom as are many other snakes but does not seek them out specifically.

Primary activity is during the day or dusk, and shaded ground is generally preferred. Six to twelve eggs are laid during the summer and hatch about seventy days later. When annoyed, it may vibrate its tail and strike viciously as several other snakes do. After gentle handling in captivity, they usually become quite docile. Many people who do not like snakes will say the kingsnake is the only good one—a viewpoint not shared by other snakes I am sure.

Poisonous Snake Bite

The idea of being bitten by some poisonous creature is foremost in the minds of some people when they think of outdoor activities. I have met people who could not enjoy camping or hiking because of the possibility of a poisonous snake bite. It is sad when a person feels this way but much much worse when they pass these fears on to their children. It is not that there isn't a possibility of a poisonous snake bite because there is a slight possibility. The chance is certainly not as great as an automobile accident on the way to a store, however, and, although the danger of an automobile accident is real, it does not deter us from the automobile.

The only poisonous snake we have in California is the rattlesnake. With proper treatment a poisonous rattlesnake bite should not cause permanent damage.

Snake venom is largely protein and many venoms have as many as 20 different proteins in them. It is a gross oversimplification to group them as neurotoxins and hemotoxins as their functions are far more varied. Experts, today, say that snake venom represents one of the highest forms of chemical evolution if not the highest. Needless to say this venom has to get into the bloodstream to have an effect. In the stomach or on the skin it is harmless.

Numerous studies have shown that in about one of five bites the snake does not inject venom. Treatment of these non-envenomated bites has given rise to bizzare and useless snake bite treatments. I am sure you have heard of some.

If envenomation has occurred, edema and swelling usually occurs at the bite within five minutes and not more than 30 minutes. Pain and change in pulse rate soon follow. Another odd symptom that usually occurs is a tingling sensation of the tongue and mouth.

If by some rare circumstance you are bitten by a rattlesnake and the above symptoms are present, above all, stay calm and go to a hospital or doctor for the antivenom treatment. Take your time and do not have an accident on the way.

Sharp-tailed Snake

Is there a snake that can sting you with its tail and poison you? No, of course not, but this is one of many myths told about snakes that perpetuate people's fear of them. I have found that in some cases individuals have made up stories about harmless snakes in order to justify their own fears. This is probably why the little sharp-tailed snake, *Contia tenuis,* is thought by some to sting.

This quite small and harmless snake may be only 8 inches long or may grow to 18 inches. The young may be only 3 inches long. The adults are rather stout and reddish-brown or gray above. They are often more reddish toward the tail. Usually, there is a rather indistinct yellowish or reddish line down each side of the back. The underside is quite distinctively marked with alternating bars of black and cream on each belly scale. The name sharp-tail is given because of the rather sharp conical spine at the tip of the tail.

This little secretive snake is almost never seen naturally in the open. It frequents woodlands, grasslands, pastures and meadows—especially those near streams or other water. I have found them under bark, logs and rocks. Locally, they are often found under and in wood-chip piles where they find slugs which are evidently their only food. Some authorities believe the pointed tail helps anchor the snake so it can pull or push the slugs loose. The long teeth of the little sharp-tailed snake make it possible to hold on and swallow the slugs.

One other interesting characteristic of the sharp-tailed snake I have noticed is its ability to climb up vertical glass. It somehow clings to the glass. I have had many mysteriously escape before I discovered this ability.

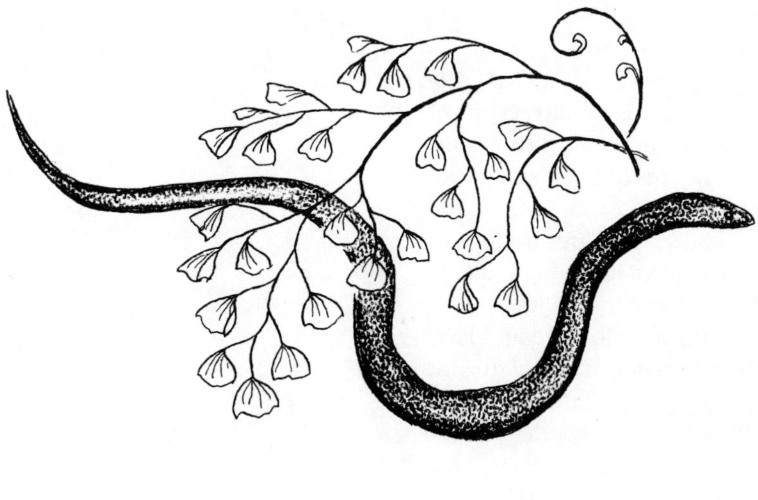

Western Racer

Even harmless snakes, make some people feel very uncomfortable, and that is too bad. Usually, if people feel this way, it is because their parents or relatives passed this feeling on to them. People attending my classes have shown me that almost anyone can overcome their fear of snakes by an intelligent approach. If you learn about snakes and have them around, it is usually easy to learn to handle them. You may then be able to lose the fears that may have kept the out-of-doors from being an enjoyable place.

One of the interesting snakes found in our area is the western racer, *Coluber constrictor mormon.* The adults are 2 1/2 to 4 1/2 feet long and are quite slender. The tail is long and pointed. Color above is an olivaceous or brownish while the underparts are a pale yellow. There is really no other snake in our area that is at all similar.

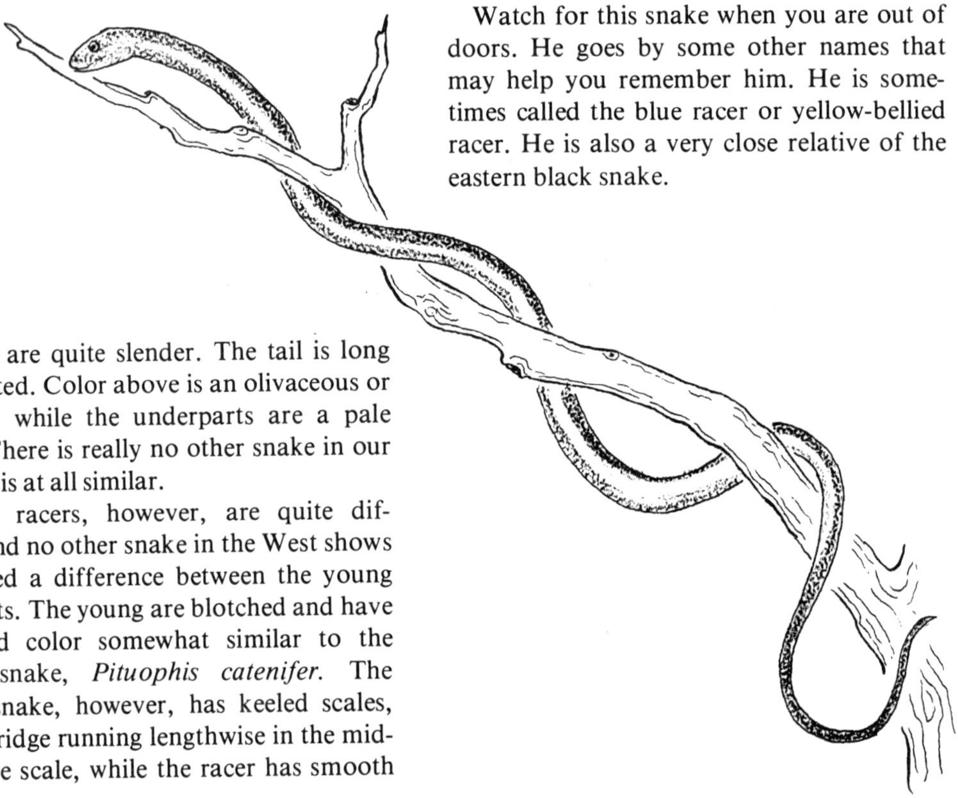

Young racers, however, are quite different, and no other snake in the West shows as marked a difference between the young and adults. The young are blotched and have a ground color somewhat similar to the gopher snake, *Pituophis catenifer.* The gopher snake, however, has keeled scales, that is a ridge running lengthwise in the middle of the scale, while the racer has smooth scales.

Racers are a very active and fast-moving snake. When hunting, they hold the head and neck well up off the ground— undoubtedly for a better view. They are also good climbers and probably take some baby birds and eggs as well as lizards, tree frogs, insects, rodents, other snakes, etc., anywhere.

When feeding, they do not constrict but may hold the prey down with a loop of the body and start to swallow. They are a very alert snake and will often bite freely. I can almost promise you a bite if you pick one up and leave his head free. Their teeth are relatively short, however, and unless your hand is jerked from his mouth, they probably won't penetrate the skin.

Watch for this snake when you are out of doors. He goes by some other names that may help you remember him. He is sometimes called the blue racer or yellow-bellied racer. He is also a very close relative of the eastern black snake.

3
Invertebrates
Worms, Myriapods, Crustaceans, Insects, Arachnids

I hope that everyone realizes how intricate our environmental relationships are, and how one small change in it can cause many undesirable events.

Earthworm

The rains of spring bring many things to the world of nature. The awakening of slumbering trees, the blooms of brilliant flowers, and blankets of lush green grass are but a few. They also cause an interesting phenomenon of earthworms.

The common earthworm, *Lumbricus terrestris,* is close to 11 inches long. Its long segmented body has a pointed head and slightly flattened tail. There are short bristles on the undersides of the body that help him move through the soil. A mucous that is secreted also helps.

Male sex organs are located on the 15th segment while the female organs are located on the 14th. Two worms mate by being head to tail and secreting sperm into each other's receptacles. After mating, a cocoon is secreted and then shed over the head. In the cocoon are its own eggs and the sperm of the other. The eggs hatch 2 or 3 weeks after fertilization takes place within the cocoon.

The earthworm, or as many call them fishing, or angle worm, burrows underground. They can literally eat their way through the soil, passing it through their digestive tracts. They digest the decaying organic matter and expel the indigestible matter in little pellets. When conditions permit, however, the earthworm feeds outside its burrow. It generally comes out at night and feeds on various decaying matter. The animal can retract into its burrow very quickly, however, and its abilities to extend itself are well-known.

The earthworm favors damp, relatively warm soil. When the surface soil is too warm or cold, they burrow deeper. The burrows of the earthworm keep the soil open for the absorption of water and air. The value of the earthworm in this way is almost impossible to estimate.

There are many species of earthworms and *Lumbricus terrestris,* is just one. Various species have been spread around the world in potted plants and by fishermen raising them for fish bait. In California they are found up to the middle altitudes of our Sierras.

Earthworms are noted for their sensitivity to soil vibrations. They can often be brought to the surface by pounding a stake into the ground or by some similar vibration.

The phenomenon I mentioned previously is the way many earthworms come out on the surface of streets and sidewalks when it rains. It is generally stated that they come out when the soil is waterlogged, and they need air. I am not convinced of this myself as I have kept an earthworm healthy in an aquarium for several months. What do you think?

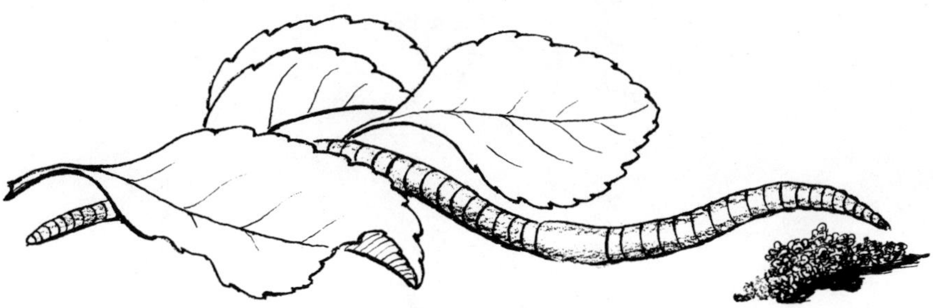

Centipedes

Summer weather and gardening bring us into contact with many creatures of the earth. One of the most unique is well-known to us as the centipede.

Centipedes belong to a group or class called Chilopode. They are not insects as insects belong to another class called Insecta. Both the insects and centipedes are in the same group called a phylum. Their phylum name is Arthropoda or jointed-legged animals. Spiders, ticks, mites, crustaceans, scorpions, and millipedes are also in the phylum Arthropoda.

Centipedes are very flat creatures that have very long, slender, wormlike bodies. They have a varying number of body segments and one pair of jointed walking legs on each of these segments. This is one of the characteristics that separates them from the millipedes which have two pairs of legs per segment. Centipedes vary in length throughout the world with the smallest being one-tenth of an inch while the largest is nearly a foot. They have one pair of rather long antennae, and some species are without eyes. The rounded head is without legs, but a pair of modified legs form the claws. The claws are used to poison their prey as centipedes are entirely carnivorous. They feed upon snails, slugs, worms, other centipedes, insects, larvae and other small animals. They hide by day and hunt by night.

Centipedes always live in damp, dark, out-of-the-way places such as under stones, fallen leaves, logs, under bark and in crevices of the soil. Because of their retiring habit, centipedes cause us few problems. Ours are not generally considered dangerous, but being bitten by larger ones may result in a very painful wound. Even the large ones in the tropics are not considered fatal. There is a large greenish one in southern California, *Scolopendra heros,* that is feared because of its poisonous bite and the fact it produces a reddish streak where it crawls on the body. Our largest centipede could probably give a bite about the same as a wasp sting.

The numerous legs of a centipede and how he controls all his legs has always fascinated us. His problems are reflected below:

I'm glad I'm not a centipede,
And here is the reason why.
Because I'd have one-hundred legs,
And one-hundred shoes to tie.

Think of that at today's prices!

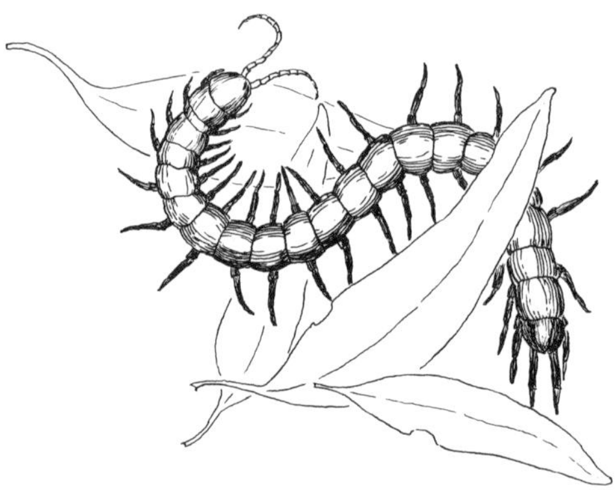

Millipedes

We find many interesting animals when we look in the small hidden places of our yards. Insects, worms, slugs, snails and spiders we may know a little bit about, but there are some that we know little more than the name. One of these is the millipede.

Millipedes belong to a group called the diplopods. This is the same level of classification as the centipedes, insects, arachnids and crustaceans. Millipedes have long wormlike bodies that are rounded instead of being flat like a centipede. The name diplopod refers to the fact that they have two pairs of legs per body segment. This large number of legs gave rise to the name millipede which translated means thousand legger, a name often used. They have a distinct head with a pair of short, unbranched antennae, at least two pairs of jaws and usually eyes. The smallest are about one-fourth of an inch, while the largest ones are six inches. Most of the largest are found in the tropics, but there are some almost as large elsewhere.

Millipedes prefer dark, damp places under decaying leaves and logs. They are also found in loose organic soil. Evidently moisture or humidity is one of the most important factors in their lives. This response to moisture has some economic importance for under dry conditions they may be forced to attack live plants for the sake of water. Millipedes are essentially vegetarian and feed on a variety of plant substances. Normally their weak mouthparts cause them to feed mostly on soft or decomposing plant tissues. In this respect they aid in the breakdown of humus. Some species, however, are well-known agricultural pests. One of our worst offenders is the spotted snake millipede, *Blaniulus gultulatus.* This species has long been known to attack sugar beets, potatoes, oats, wheat and strawberries.

Millipedes are occasionally eaten by many animals. Toads evidently will eat any they find and the imported European starling may use millipedes for a bulk of their diet when available. Most animals, however, seem to find the millipede somewhat unpalatable because of their tough shell and gland secretion from near the legs. This secretion may include hydrocyanic acid, iodine, quinine and chlorine. A combination like that should discourage anything. If you want to see what I mean, put one in a jar, shake it, and take a smell. Whew!

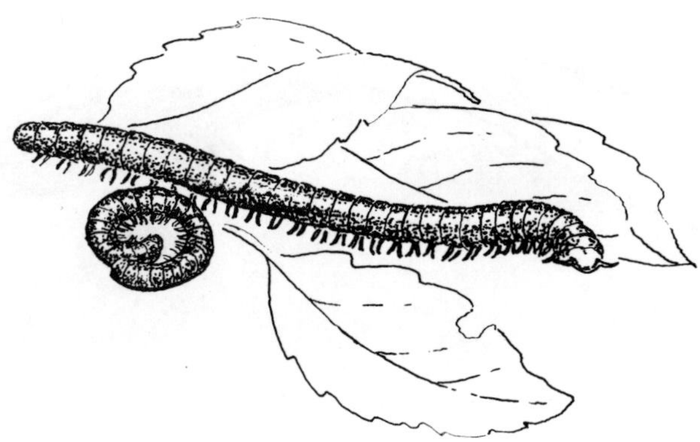

Sow Bug and Pill Bug

When nice weather comes to our valley people find themselves outside gardening and cleaning up after the long winter months. Picking up boards, papers, rags, etc., leads to the discovery of many interesting creatures of nature. Many of them we see year after year but know nothing about them.

Two of the tiniest of our backyard residents are the dooryard sow bug, *Porcellio loevis,* and the common pill bug, *Armadillidium vulgare.* These arthropods are not insects but crustaceans, and most crustaceans are found in water like the crayfish, crab and shrimp. The sow bugs and pill bugs, however, are unique dryland representatives. They belong to a group called the isopods of which most are small marine organisms. The term Isopoda means equal feet.

Both the sow bug and pill bug are about three-fifths of an inch long and one-third of an inch wide. They are a deep slate color with two pale indefinite longitudinal lines. The underside is more pale. The back or covering of the body, the exoskeleton, is composed of a series of articulating plates that give these animals an armored appearance.

The seven pairs of legs are barely visible from the back, but the two antennae are nearly half the length of the body and quite obvious. The young are very much like the adults but are smaller and more pale. The eggs and young are carried in special pouches under the female's body.

The pill bug can be told easily from the sow bug by its ability to roll up when disturbed. The sow bug does not do this.

Both animals are strictly nocturnal, and during the day are found under stones, boards, rotting wood, bark, and other similar places. They are mostly scavengers although they will eat tender young plants and can be a pest in greenhouses and gardens.

Moisture is one of the critical factors for these crustaceans. They breathe by gills on their legs which must be kept moist. This probably is why they are more active in wet weather.

Watch for them and don't hesitate to pick them up. Naturally, they are harmless.

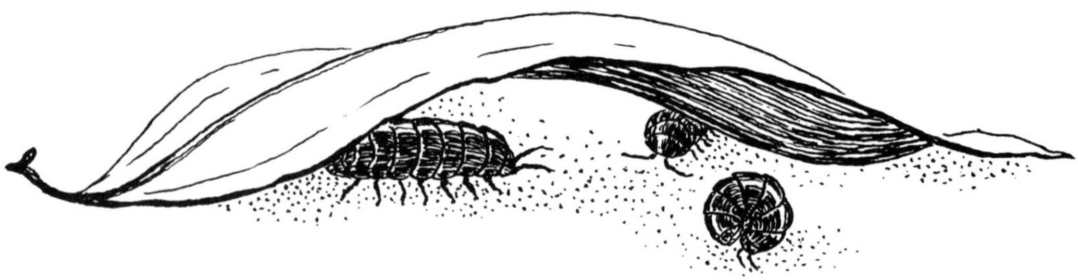

Darkling Ground Beetles

Many creatures of nature nearly escape notice by the casual observer. Many birds, mammals and insects draw much attention to themselves while others go about their business scarcely noticed. One of these little-noticed insects is a carabid beetle called the predaceous or carnivorous ground beetle. This large family is usually dark and seldom metallic in coloration. They can be small or large and are nocturnal in habits, hiding under stones and logs during the day. They can be recognized by the way they scurry around when they are disturbed. They may or may not have well-developed wings, and many secrete a malodorous and volatile fluid when disturbed. This may happen if they are handled.

The well-developed mandibles or mouthparts are one of the characteristics of this beetle. They are not furrowed or grooved as are the mandibles of some other beetles, but they are powerful. They are used to subdue and crush their prey—their food being caterpillars, grubs, grasshoppers, snails, and other small nocturnal animals.

The eggs are usually laid in the ground and the larvae are fusiform or tapered at both ends. The adults have antennae that are 14 jointed as are all beetles. This family, as a whole, is one of the most beneficial groups of insects known because of its destruction of other pests.

There are several species of these beetles found in our valley. The snail eaters belong to the genus *Scaphinotus* and are large black or reddish-brown beetles. Another genus known as *Calosoma* are large black, green or bronze beetles. One of these is the common black calosoma, *Calosoma semiloeve.* This one-inch long beetle is black, and almost smooth. The most common species in California is *Pterostichus californicus.* This species is common under stones, and is often found feeding on the larvae of the codling moth.

Watch for these environmental partners and help them if you can.

Mourning Cloak Butterfly

The signs of spring are many and varied to the observers through nature's window. The grass grows quickly, flowers appear, mosquitos annoy us, and we see butterflies flitting across the meadows. It is unusual to see signs of spring in the middle of winter, but it does happen. One of the most interesting false prophets of spring is the mourning cloak butterfly, *Nympholis antiopa.*

This common butterfly is familiar to almost everyone. The wing edges are irregular, but the hind edge of the front wing is very straight. The hind wing has a short tail that is suggestive of a swallow tail. The wing span is about 2 1/2 inches with the outside edges broadly margined with yellow. The yellow is bordered by a parallel row of bluish spots. The basic color of the top of the wings is dark brown to black while the undersurface is dusky.

The larvae or caterpillar is 2 inches long with an angled head. The body is black with many forked spines and whitish spots. The middle of the back is red. Groups of the caterpillars are usually found together on the leaves of trees like willow, cottonwood and aspen.

The one-inch pupa or chrysalis is dark brown with red-tipped swellings. It is often covered with a whitish powder. The pupa stage may last from 8-16 days. The eggs are dark orange-brown or black and are laid on twigs not leaves. This is just one of the many unusual characteristics of this butterfly.

The name mourning cloak comes from its dark color that is similar to clothes worn in mourning. Adults occur throughout the year but may hibernate in winter. They will, however, appear during the sunny days of February and March, often being drawn to the nectar of pussy willows.

Keep your eyes open for this unique, pretty butterfly. It can be a pleasant taste of summer in the midst of winter.

Monarch Butterfly

To talk of butterflies in the fall would seem almost to be after the fact. We usually associate butterflies with spring, but such is not necessarily the case with the monarch.

The monarch, *Danaus plexippus,* is very obvious in the fall because of an influx from the north. The life history of the monarch is more complex than most other butterflies because it migrates.

The adult monarch, sometimes called the milkweed butterfly, has reddish-brown wings bordered by black. The black margin has two rows of small white spots throughout most of it. The wings also have black veins. The caterpillar is banded with black over a general yellowish-green color. There are two threadlike appendages at each end of its body.

The chrysalis or pupa when first formed is a beautiful pale green with glittering gold spots. It fades later. The eggs are green and shaded like flasks with lines of tiny pits. They are laid only on milkweed leaves and hatch in about four days. The caterpillar pupates in about ten days to the pupal stage which may last about ten days. Both the larva and the adults seem to be protected by a bitter taste, probably because of the body juices. They are seldom bothered by predators.

The migration tendencies of the monarch are greatly misunderstood. Large numbers fly south from the northern latitudes each fall, and yet they seem to fly north in the spring. The ones flying north, however, are not the same ones that flew south. The longest flight recorded by a tagged adult monarch is from Ontario to Mexico, a distance of over 1,800 miles. The adults may reproduce on their wintering grounds or after a short flight northward in the spring.

There have been many explanations for occurrences like the butterfly trees of the Monterey area. One of the most plausible is that in their flight south, many of these breezes blow ashore at Monterey and perhaps the exhausted butterflies rest on certain trees.

In the fall, listen and watch for word of when the butterflies start arriving on their trees. A visit to them is an unforgettable sight.

Crane Fly

As the weather warms with the coming of spring, the desire of people to spend their time outdoors increases.

This venture outdoors, however, is not always as pleasant as one would like because of various pests that also increase with the arrival of spring.

Everyone is well aware of the discomforting abilities of the mosquito. They have sent many a brave soul indoors from patio parties and other outdoor activities. The whine of a mosquito is so well-known that most of us start looking when the whine stops, and we seem to slap by instinct.

This article is not, however, about a mosquito but about an insect that is often killed because he resembles the mosquito. This insect is called the crane fly, and he does not buzz or bite. He is often seen in the same areas as mosquitos but is not nearly as agile or clever in escaping our slap, so often is killed by mistake.

The giant crane fly is one of the most common found in the West, The body is from 1 to 1 1/2 inches long, and the wingspread may measure 3 inches. The color is an even reddish-brown although the thorax (the segment with the wings attached) may have white markings.

All in all, it looks like a giant mosquito. The legs are very long and easily torn from the body. The larvae are about 1 or 2 inches long and dull brown in color. They live in moist situations or grass roots, leaf mold, trash or mud.

The crane fly is closely related to the mosquito. Both are in a group called Diptera which means two wings. They cannot puncture skin as do the mosquitos, and so they are harmless to us.

They are sometimes called mosquito hawks since some are thought to kill other insects. Although they resemble mosquitos, they are different. Can you tell them apart?

House Cricket

In our technological world of today it would seem that man could easily outwit the simple creatures of nature. We build a house and cover it with several layers of various materials to exclude wind, rain, and we think everything else. The windows are sealed, the doors weather stripped, and screens placed over all entrances in a futile attempt to protect man's domain from nature. Most of us understand how flies and mosquitos infiltrate these precautions, and a close investigation usually reveals how the tiny ant finds his way inside.

It is amazing, however, that the lowly house cricket is able to completely outwit all the precautions man takes. Who has not turned on the lights at night and seen a cricket scamper across the floor or moved a box from a corner and found a cricket crouched behind?

Finding a cricket in the house can mean different things in different countries. In Malay, it is believed that a cricket is the assumed form of the spirit of the dead. In China, cricket fighting is as popular as cock fighting in other parts of the world. The Cherokee Indians call the cricket "talatu" or barber, and in Virginia and Maryland their presence is an omen of prosperity.

The house cricket, as we generally call him, is not the true European species but one of the field variety. They often seek the shelter and warmth of human habitation and can be a pest. When in the house they may feed on woollens and food products and, outside, eat grass, grains, or fruit. They are about one inch long and have a color from light brown to black. From their head there are two long threadlike antenna and the hind legs are very long for jumping.

Crickets are said to be the most musical of all insects; however, only the males chirp, or stridulate. Their chirping is done by scraping a filelike structure beneath one wing over the opposite wing that has a roughened top surface. The wings themselves act as a sounding board. The female catches this symphony through an ear located on her foreleg. It can be truly said that a cricket sings with its wings and hears with its elbows.

The frequency of chirping increases with the temperature and if you count the number of chirps in 15 seconds and add 38, you will approximate the temperature. Don't take my word for it, however; try it yourself.

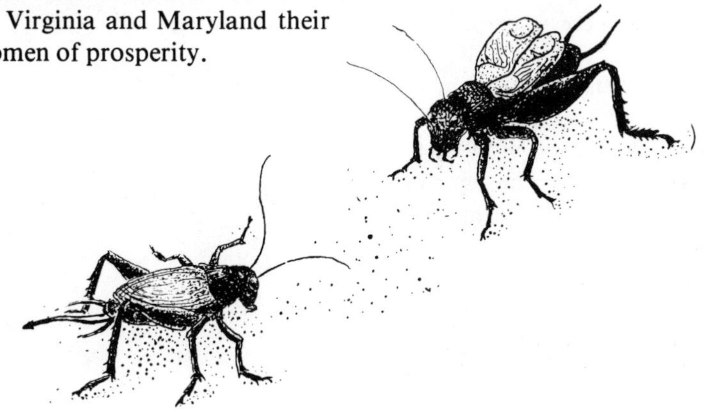

Dragon Fly

I was walking across our campus at American River College one warm summer day and noticed several dragon flies flitting over the open area of lawns. One usually expects to find them near water and since our lawns were just watered, they were probably attracted by the other insects that were flying

For many viewers through nature's window most insects are just "bugs," but dragon flies are recognized by everyone. The superstitions and folk tales concerning them, however, often cause people to be afraid and not appreciate what this wonderous insect does for us.

By the uninformed, dragon flies are thought to feed and doctor snakes; sew ears, lips, fingers, nostrils or eyelids together; affect fishing luck and even sting. All of these feats are far beyond the powers of any insect. Some of the names given to this creature seem to promote these false ideas. They are often called darning needles, snake doctors, snake feeders, horse stingers, water maidens and mosquito hawks.

The term mosquito hawk is very apt because the dragon fly has a voracious appetite for living insects. Only the praying mantis may eat more than the dragon fly. It has been estimated that the dragon fly may eat his own weight in insects every half hour. They have been captured with more than a hundred mosquitos in their jaws.

Usually prey is scooped up by the legs that are held basketlike when flying. The legs are bunched at the front of the thorax and are not used for walking, only for catching prey and perching. In the insect world the dragon fly is unrivaled in aerial grace and speed.

Dragon flies are members of the insect order Odonata which means, those with a tooth. The term tooth refers to the tusklike abdomen. The colors of this order are often brilliant and may range the hues of the rainbow. The eggs are usually laid in water and the young may spend several years as voracious aquatic nymphs. As adults they last only the season and usually die with the first frost.

Watch for this fascinating ecological partner. He makes your life more pleasant.

European Earwigs

It is not unusual for the obvious creatures of nature to be recognized by almost everyone. They need not be large or cause pain to be well-known but can be numerous. Such is the case of the earwig.

Earwigs, as a group, are slender insects that have large forceps—like appendages extending from the rear of their bodies. The adults usually have four wings, the front pair being short and leathery while the hind pair is membranous. The texture of the forewings is responsible for the scientific group or order name of Dermaptera. This means skin wings.

The name earwig evidentally had its origin in a very old superstition that the insect would enter people's ears. Although the idea seems logical, there seems to be absolutely no substantiation that this ever occurs. The name is derived from an old Anglo-Saxon word earewicga (eare = ear + wicga = beetle or worm).

It is quite obvious that this insect is essentially nocturnal in habit. During the day they hide in cracks, crevices, under bark, in flowers, holes or any other suitable dark place.

There are several kinds of earwigs found in the United States, but the only one of importance in our area is the European earwig, *Forficula auricularia.* This common insect is about 3/4 of an inch long and dark brown in color. It was introduced accidentally from Europe and spread rapidly. It appeared in Rhode Island in 1912 and Berkeley in 1923.

Young and adults come forth at night and attack plants, including flowers, ornamentals fruit trees, vegetables, grains and grasses. They may even become a pest in houses. It is probably not how much they eat that makes them unpopular but their mere presence is obnoxious.

The function of their pincers on the tail end is not too well understood, but they are probably used in defense. The male has bowed pincers, and the larger ones might even be able to pinch you. They do not bite, however, and are easily handled. Try it—you may find it fun.

Oak Galls

The ashy-gray skies of our valley often provide a striking backdrop for the many trees that lose their leaves in winter. Their silhouettes are as varied and fascinating as the clouds in the sky. One of the most obvious aspects of the leafless oaks is the balls that are seen hanging from the branches. I can't help but smile when people ask, without thinking, if the balls are the fruit of the oak when, of course, they know the fruit of the oak is the acorn.

The balls you see on the oak are the result of the eggs of a tiny wasp in the family Cynipidae. Various species of the genus Amphibolips lay their eggs under the bark of a twig. The eggs hatch into larvae. The larvae feed on sugar that is formed when enzymes secreted by the larvae convert the carbohydrates of the stem into sugar. It is thought this increase of sugar causes that part of the plant to grow rapidly forming the gall. In other words, the gall grows with the larvae. These galls or oak apples as they are often called can be several inches across and may grow rapidly. They usually start growing in the spring, and the adults may emerge in the fall. When the gall is cut while it is green it turns purple or black very quickly. This is due to the large amount of tannic acid present. Often the larvae can be seen when the gall is cut. It has been estimated that more than 500,000 wasps were living at the expense of one oak tree at one time. That is quite a family.

There is very little economic importance attached to our oak galls. They probably damage the host tree very slightly. Galls from oaks in Europe, however, have long provided the best of permanent inks. This gall-made ink has been specified for use by the United States Treasury, the Bank of England, and others. Galls are often an important source of tannins used in tanning leather. In Missouri and Arkansas a type of gall has been used as a food for domestic animals. Many galls have been used in medicines, and the ancients had many superstitions about them. As far as I can find out, however, our galls are not used for anything, except perhaps fishing bobbers.

Take a look at this mini ecological unit the next chance you get. It has quite an interesting story as you can see.

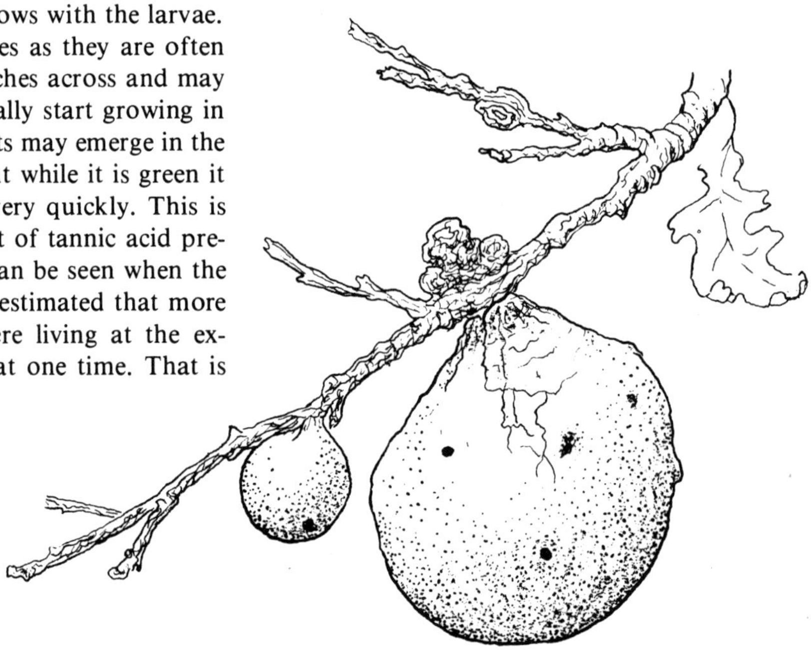

Jerusalem Cricket

When spring comes, various things happen to a person's outlook. As the old cliche says, "In spring a young man's fancy lightly turns to thoughts of love." To the home gardener, however, spring is a time for cold muddy hands and a sore back. It is also a time for the discovery of one of the most peculiar creatures of nature.

Gardeners often find these mini monsters in the earth, under boards, rocks, and clumps of weeds. Their ferocious appearance will always startle you. It is difficult to believe that anything so fierce looking is not poisonous. They are, however, very closely related to the grasshoppers and locusts.

Perhaps you have already guessed I am referring to the harmless Jerusalem cricket, *Stenopelonatus fuscus.* This ground dweller is up to 2 inches in length with a pale amber head, chest and legs. The bulgy brownish abdomen is very striking with the wide almost black bands running across it. I suppose, however, that it is the large, almost human looking, head that causes the major fear. The huge jaws of this cricket work from side to side and must be considered with respect. They can deliver a good solid pinch but cannot penetrate the skin.

The Jerusalem cricket or as it is sometimes known "child of the earth" is essentially nocturnal in behavior. At night it roams freely in search of dead animal matter, roots and tubers. Perhaps that is why many call it the potato bug. They are also fond of other insects and even indulge in cannabalism. The female frequently devours the male after mating.

In spite of their great strength, powerful jaws, and spiny legs, they are often prey for other creatures. Their remains are often found in the stomachs of pallid bats, and I have seen a captive black-headed grosbeak devour them with great gusto.

If you find one, grit your teeth and pick it up. It can't hurt you, and you will find a truer understanding of the world of nature. It will also do a lot for your ego. Try it and see.

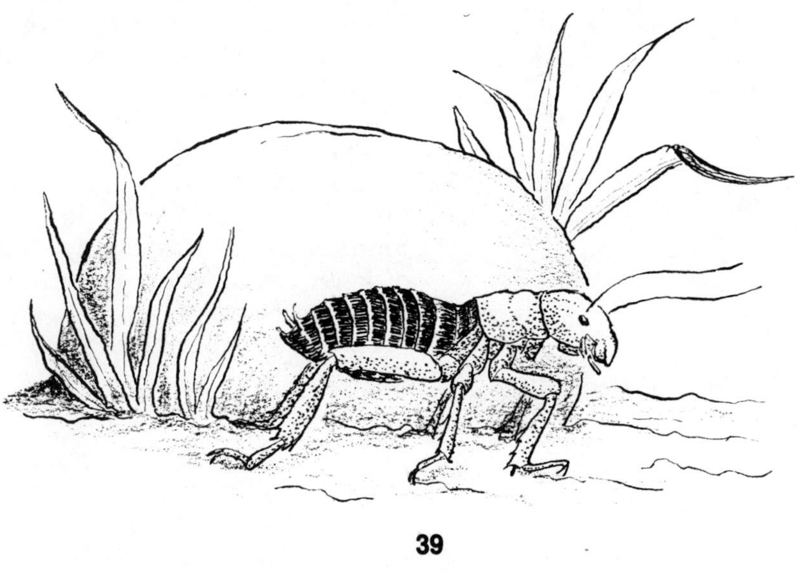

Lacewing

Warm summer evenings bring many creatures to the attention of those who look through nature's window. Mosquitoes, house flies, moths, crickets and frogs are just a few. The porchlight, lighted store window, or other sources of light at night bring in many fascinating insects. None, however, is more interesting than the California green lacewing, *Chrysopa californica*.

This common western species is pale or bright green with yellow or golden eyes and a yellow stripe on its back. As in all Neuropterans, nerve wings, the large transparent lacelike wings are held rooflike over the back and abdomen. Their long, threadlike pair of antennae can be directed forward or backward. They are rather clumsy fliers and easy to catch.

The female lacewing when laying her eggs first touches the tip of her abdomen to a leaf or stem. She secretes a pool of a gluelike secretion, and then, with a quick movement, raises her abdomen so that a hairlike stalk is formed. She then deposits an egg on the tip of the stalk. She moves a short distance and repeats the operation until a group of eggs is laid. They look like a row of miniature lollipops.

In about five days the egg hatches, and the lizardlike larva climbs down the stem and starts its search for food. The voracious little predator would eat the eggs of his brothers and sisters were they not on the top of a stalk he could not climb up. The favorite food of the larva is aphids, however, and they have the ability to eat almost one a minute. The four jaws are shaped like calipers but have grooves. These pierce the prey, and when the prey is raised, its juices run down the grooves into the mouth of the larva. In the larval stage the lacewing is often called the aphis lion—a very appropriate name.

When enough aphids are consumed, the aphis lion attaches itself to the underside of a leaf. It then draws its body up into a ball and spins a white coverlet of silk about the size of a pea. In little more than ten days, the adult emerges from the cocoon to start the cycle again. The adult can give off a very nauseating fluid when disturbed. This is undoubtedly a protective device.

Light traps that draw insects in and kill then are a serious threat to this completely beneficial insect. Undoubtedly, this method of insect control causes an increase in the aphis population. As a rose fancier, I say, "Hooray for the lacewing."

Ladybird Beetle

One of the fascinating things I have learned in my years of teaching is the effect that a name has on people's attitude toward various creatures. The names gila monster, butcher bird, and the like evoke feelings of resentment or disgust, while names like daddy longlegs and ladybug, do just the opposite.

The ladybug or as it is more correctly called ladybird beetle, is an insect that almost everyone knows about and likes.

The most common one in California is the California ladybird beetle, *Coccinella californica.* It does not have the rather typical black marks on the elytra or wings. Instead, it has a plain red back with black head. The ladybirds are all beetles or Coleopternans, however, and not bugs which are in a different group called Hemipterans.

The ladybird beetles are all in one family Coccinellidae, and all but one small group are beneficial to man. The beneficial ones are predaceous and feed on other harmful insects. They are about one-third of an inch long and may be red, black or yellow. They attack numerous scale insects, aphids, and the eggs of many insects. Both the adults and larvae feed on other insects and are considered one of the most beneficial groups of insects known.

The larvae are blackish grubs that are often spotted or banded with red or yellow and armed with spines. The oval-shaped body has distinct segments that give it an accordianlike appearance.

The eggs are laid in spring and hatch shortly thereafter. The larvae begin eating right away and grow rather fast. Eventually they pupate and hang by their tails for a few days till they transform into adults. As adults they start eating insects immediately and continue through their lifetime.

The convergent ladybird beetle, *Hippodamea convergens,* is often found in large numbers in certain warm spots in canyons of rivers and creeks in the fall. They hibernate in these accumulations till spring. Then they disperse over wide areas and produce their progeny.

I hope you all enjoy these colorful little pest exterminators as much as I do.

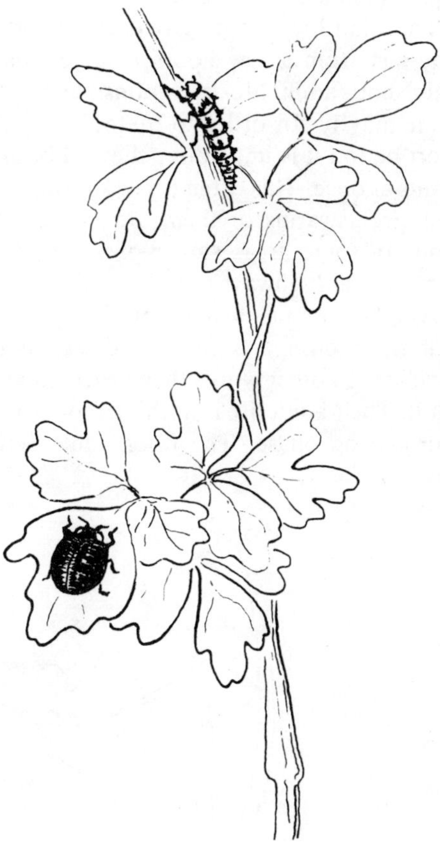

Mayflies

There are several phenomena in the world of nature but none more unusual than the life of the mayflies. This group of insects has one of the most unique life cycles in the animal world. They are very delicate creatures that are abundant around or in streams, ponds, and lakes. Adults almost all have two pairs of wings with the front pair being much larger than the hind pair. Two or three tails or cerci are present, and the front pair of legs are held conspicuously far out in front. The well-veined gossamer wings make flight possible but not strong.

The adult life of a mayfly is very short. It may last from a few hours to a few days. This short adult life is responsible for the name mayfly. In order to mate during this short period it is imperative they all become adults at once. This often occurs during May and large swarms of them are not uncommon. In some areas snowplows have been used to clear highways of their bodies.

Another unique feature of this group is that the mouthparts of the adults do not function. In other words they never eat as an adult. Their adult life is nothing more than a reproducing phase. The brief adult period caused the group to be named Ephemeroptera derived from the word ephemeral. We often hear the word ephemeral used to describe a woman's beauty. The user undoubtedly means to imply delicacy but ephemeral means "lasting but a day." That is hardly a compliment I would say.

The eggs are deposited on the surface of the water where they soon settle to the bottom. They soon hatch into naiads (aquatic nymphs) and stay in this stage up to three years. They will molt many times as they grow larger. Most eat dead or living plants and they, themselves, provide much food for fish. The naiads are a good indicator of pollution since they must have well oxygenated water. When the first adult stage is reached in the water, they are called "duns" by trout fisherman. In a few hours they molt again and acquire the adult coloration. The second molt of the winged stage of an insect is unique to this group.

Look for mayflies near the water in the daytime and around lights at night. This odd character will probably visit you in summer. Watch for him.

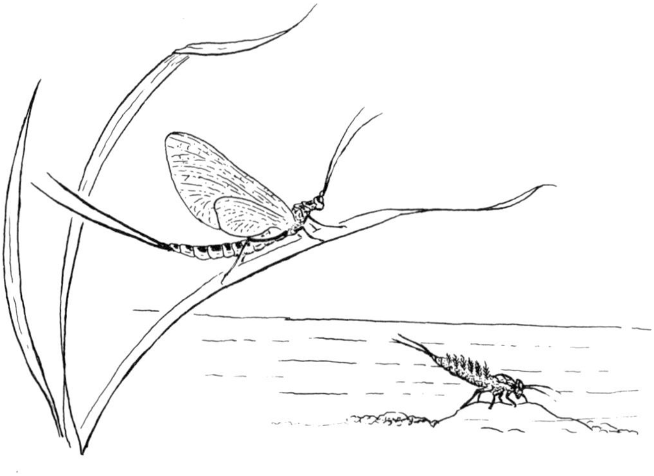

Praying Mantis

Praying mantis, the soothsayer, the nun or the devil's rearhorse are all names for one of the most interesting and popular insects we have. Praying mantis is probably the most commonly used name.

The number of species we have in the United States varies from one authority to another. One writer says there are four, another 14, and one indicates there are 10 found in just the West.

The ability of the mantid to remain motionless in its typical praying pose makes him hard to find. Nevertheless, we do see them in summer when they develop wings and move around.

Fall is the end of the life span for the mantids, and they all die when cold weather approaches. They will die even if kept warm and well fed.

Late summer and fall finds the mantids full grown and sexually mature. The female often kills and eats the male after mating. She can be distinguished from the male by her heavier abdomen.

Soon after mating she secretes a frothy white mass about the size of a shelled walnut. She then deposits her eggs in rows in individual chambers in the froth. The froth soon hardens and darkens to a rootbeer color. The eggs overwinter in this stage. When spring arrives and the temperature rises, the egg mass softens and soon the young mantids emerge. They are predacious on other insects and live a solitary life looking for food on their own.

The young capture prey the same as the adults. Their forelegs are lifted in a reverent manner as if they are praying. They remain in this pose or may move forward cautiously until the victim is within reach. At this time the forelegs dart out and clamp shut over the back of the prey. The parts of the legs closing over the victim have spikes facing each other. The helpless victim is held in a very effective vice.

I am very glad praying mantids are not the size of cows.

Stinging Insects

The whine of the wings of a stinging insect often turns even the bravest of outdoor enthusiasts into a cautious coward. Even though these stinging members of the order Hymenoptera are all small and essentially harmless, their numbers and unpredictable behavior are difficult for us to cope with. Few people pass through childhood without having been stung by at least one of them.

The ground-nesting bumblebee of the genus Bombus is one of the most conspicuous of the stinging insects and can deliver a painful sting. Children with bare feet sometimes tramp on the bee itself or come too near the opening of the nest. Usually, a bumblebee or two guard the opening and will attack if aroused.

The numerous honey bees also sting many barefoot children every summer. The white dutch clover of our lawns and parks attracts the bees. They are also attracted to people by the use of their hair tonics, perfumes and sweets.

The yellow jacket, *Vespula maculifroms,* is a paper wasp that hides its nest in the ground. The sting is especially miserable as is the sting of its close relative the bald-faced hornet, *Vespula maculata.* The bald-face hornet is largely black with yellowish-white markings. The yellow jacket has much more yellow. The nest of the bald-faced hornet is a basket-shaped bag with a hole at the bottom. The nest is often basketball size and is usually attached to a tree branch. The hornet is quite vicious in defense of this nest and an intruder may be attacked and chased by hundreds of the hornets. Believe me, it is best to avoid a nest. If the nest is near your house and creates a hazard, any common spray pesticide can be sprayed in the opening at night. Sevin is particularly effective. Do not molest them during the day.

It is best to avoid stinging insects, and if one is near you, stand still or walk away. Do not swat at them. If you are stung, a drop of amnonia or an ice cube on the sting may help.

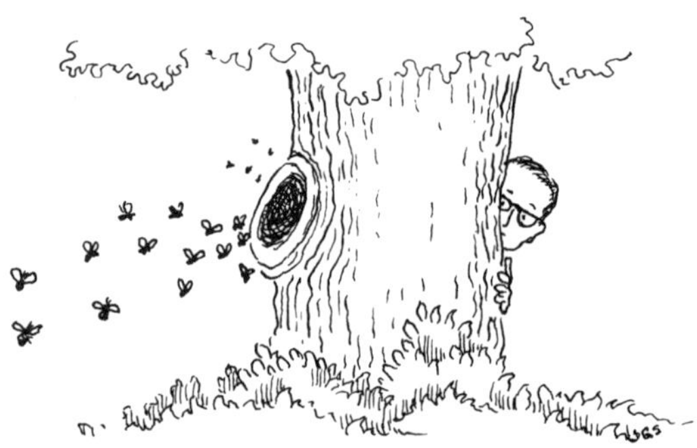

Tent Caterpillar

The art of watching nature has many facets. Sometimes we use binoculars or cameras to explore it. Othertimes, we use the unaided eye to look closely at nature's minutia. We are fascinated by the working of an ant colony or by the antics of a spider. Few people, however, have observed the organization of the tent caterpillar, *Malacosoma disstria* or *Malacosoma californica,* but their life cycle is quite interesting.

The home of the tent caterpillar is composed of a series of sheets of silky threads. These are secreted from the mouths of certain individuals in the colonies as they move about to feed. The tent, as their home is called, is familar to almost everyone. Colonies are especially prevalent on willows, but they may also be found on cottonwoods, oaks and some fruit trees.

The gossamerlike tent that would seem to be spun haphazardly is a remarkably effective shelter against the climate. When the humidity is low outside, moisture escape from the inside is greatly retarded. The same walls are almost impervious to rain. The prime advantage of the tent, however, is its greenhouse function of temperature increase. Since the insects are cold-blooded, a warm place to raise their body temperature is quite beneficial. The young caterpillars have an uncanny ability to anticipate weather changes. Even before the weather turns hot, chilly, or rainy, they can be seen scurrying back to the tent. After their third molt, however, the bright sun does not send them back to the tent. Instead, the forepart of the body is lifted until the head is pointed directly at the sun. This reduces the surface of the body exposed to the sun's rays and saves the energy that would have been used going back to the tent.

As the time for the caterpillars to pupate approaches, they leave the trees for the earth. At this time hordes of them may be seen on the ground. They often climb railroad tracks and travel upon the rails. In enough numbers, they can stall a train on an up-hill grade by making the rails slippery.

The next time you see a colony of tent caterpillars, take time and watch. I have mentioned only a few of their antics. I am sure you will see some others.

Tomato Hornworm

It is interesting to observe that many of the most fierce-appearing or fierce-acting creatures in nature are essentially harmless to us. The violent antics of a small harmless gopher snake may actually intimidate many people. Brightly-colored or large insects often have the same effect on individuals brought up to fear creatures of nature. One of these ferocious appearing but harmless creatures is a very common garden pest. It is found on tomato plants and is called the tomato hornworm, *Manduca quinquemaculata.*

Although it is often called the tomato hornworn, it is actually a moth and the hornworm name comes from the caterpillar stage. Many people recognize the hornworm, but few recognize the adult noth. It is a typical member of the sphinx moth family with a wingspread of about 4 1/2 inches. The overall color is a streaked ashy-gray brightened by 5 pairs of yellow spots on the conical-shaped body. The wings are narrow and pointed, and the tongue is very long. The antennae of the male is thicker and rougher than the female.

The adults are usually seen, however, about deep-tubed flowers on summer evenings. They are strong fliers and have a strong wingbeat. They feed much as hummingbirds do and are sometimes called the humingbird moth.

The female lays her large shiny green spherical eggs singly on the leaves of tomato, potato, or tobacco plants. In a short time, they hatch into a green larvae that can grow very quickly. The stout caterpillar has a row of 7 shaded, slanting stripes on each side with an equal number of horizontal stripes that converge into a V. The name hornworm comes from the large horn or spine located at the top rear. It is not poisonous as is sometimes rumored. The spine, the stripes, and a few spots give him a very ferocious appearance. They may stay as a larva up to 4 weeks when they crawl 2 to 4 inches into the ground and pupate.

The pupa is about 2 inches long with a "jug handle" on the top. The pupa case is brown and papery instead of silky as are the pupa of most other moths. It overwinters in this stage, or if conditions permit, will emerge the same summer making two generations per year possible.

Put one in a jar, feed it fresh tomato leaves and watch it eat and grow. Believe me, it even eats and grows more than my two teenage boys.

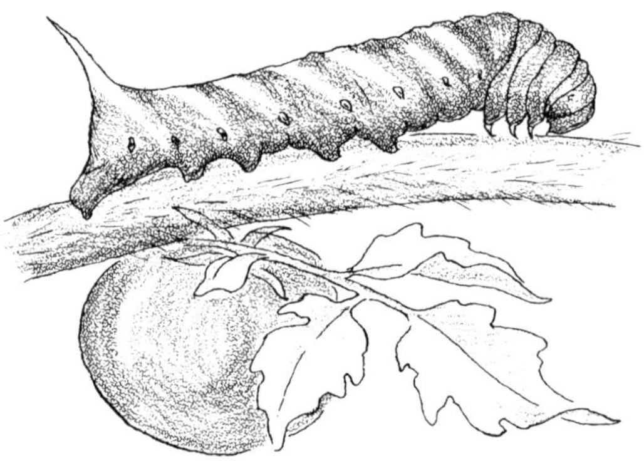

Daddy Longlegs

I always find it interesting to watch the effect names have on people. There is no question that certain kinds of names stimulate fear, anxiety, or disgust while other kinds of names promote confidence, familiarity, and even affection. People are fascinated by such names as gila monster, dragon of komodo, and Tasmanian devil. Often super powers are attributed to animals with names like these even though the animals themselves may be nothing out of the ordinary. Sometimes it also works the other way, and people that would normally fear an animal are given confidence by their friendly name. A good example of this is the daddy longlegs or harvestman.

This creature belongs to a group called Phalangida and is an arachnid like the spiders, ticks, mites and scorpions. They are common in most parts of the country and are easily recognized by their oval compact body. Their eight legs are extremely long, slender and fragile appearing. Although their body may be only 1/4 of an inch long, their legs may have a spread of 3 inches. The abdomen is segmented which makes them look different than spiders, and most of them have effective stink glands.

Daddy longlegs frequent fields, meadows, yards, leaves, logs and similar types of habitat near water. No nest or web is built, and in cold climates most die in winter. Eggs are laid in the fall, however, so the species survives. There are about sixty American species divided into two families. The appearance of large numbers of adults in the fall (around harvest time) gave rise to the name harvestmen.

Food is primarily plant juices or dead insects although some species feed on living insects. In captivity they are known to eat almost anything and are easy to keep.

In the past children were often told the legend that if their cows had wandered away, they should find a daddy longlegs and he would point the way to them. Who knows? It may have worked, especially if the children had to look all over to find a daddy longlegs.

Watch for this fellow with the friendly name, I guarantee he will live up to it.

Black Widow Spider

There are very few things in our valley of which observers through nature's window must beware. One of them that is quite common is the black widow spider, *Latrodectus mactans.*

The female, as most people know, is the dangerous one. The adult female is shiny black with a body about 1/2 inch long. A beautiful red hourglass is usually present on her underside. Her front and hind pair of legs are longer and more slender than the two middle pairs. Yes, spiders have 4 pairs of legs as compared to the insects' 3 pairs. Spiders also have no antennae while most insects have one pair. Younger females are smaller and usually orangish with angular white lines on their abdomens.

The male is about 1/2 the size of the female and harmless. As with many other species he is sometimes eaten by the female after mating, especially if she is hungry.

The web of the black widow is irregular, tough and sticky. The center is rather funnel-shaped. The eggs are laid in white or tan sacks about 1/2 inch in diameter and suspended in the web. I have noticed great aggressiveness on the part of some females while guarding their egg sacks. She is not ferocious normally. This fact undoubtedly saves us from many painful bites.

The black widow is found throughout the lower elevations of the United States but is much more common in the South and West. She likes to frequent trash piles, dark corners, holes, cellars and the like. It is thought she can bite through only soft skin, but who wants to take a chance. Her venom is more potent drop for drop than the rattlesnake's. The symptoms of her bite are chills, nausea, pain, hypertension, breathing difficulties, and muscle cramps. Of 1,000 bites reported in the United States each year, only 4 or 5 are fatal. Bites should be treated by a physician as an antivenin was perfected 25 years ago. The bite could be especially dangerous to children.

To prevent being bitten, we should always watch where we put our hands and keep them out of places we can't see. Teachers and parents should not allow the widow to stay in jars with holes in the lid in their schoolrooms or houses. They may be forgotten and might lay eggs. The young can hatch and escape through the holes and develop into adults in the schoolroom or house. Let's not increase the odds of being bitten.

Tarantula

A scary, hairy spider is an appropriate description of our largest spider, the tarantula. Its scary looks are not, however, an indication of its personality. For all its creepy crawler appearance, it is quite harmless and even makes an interesting and different pet.

Our tarantula belongs to the genus Aphonapelma and has a body length of about 2 1/2 inches. The eight legs spread out about four or five inches. They feed mainly on insects and hunt primarily at night. Their days are normally spent in an underground burrow, but they are often seen along roads at night when car headlights illuminate them

The name tarantula has an interesting background.A spider, *Lycosa narbonensis,* of southern Europe was called the tarantula, and its bite was thought to be fatal unless the victim danced furiously. The name of the dance was the tarantela which was applied for the Italian city of Tarento where the dance originated.

When frightened by an enemy, the tarantula will often rear up on its two hind pairs of legs and present its very prominent fangs. Although they contain venom, it is harmless to man, and the bite, although painful, is not dangerous. This makes them easy to keep as pets although some people have a reaction to the tarantula hairs if they handle them.

Tarantulas have been known to live up to 20 years in captivity. They will eat many kinds of insects and can go for extended periods without food. The tops of their cages should be covered because they are real escape artists. They are also cannabalistic and must be kept by themselves. The next time you see one, take a close look, and look beyond his scary exterior.

Wolf Spider

There are a few of the creatures of nature that cause some people to shudder and shake. One of the groups that evoke this fear is the spiders. Of the tremendous diversity in spiders in our area, only the black widow must be considered dangerous. There are many others, however, that are fascinating and conpletely harmless.

One of the most common harmless spiders that we see is the wolf spider, *Lycosa sp.* This member of the Lycosidae family is closely related to the water and turret spiders.

These brownish-gray spiders have the four pairs of legs that all spiders have and, of course, have no antennae. The sketch may make it appear that there are five pairs of legs, but this is not true. The first appendages that look like legs are not. They are attached to the head not the thorax as are the legs. They are called pedipalps and serve the spider as feelers.

Wolf spiders are all very hairy with long hairy legs. There is usually a couple of light-colored curved stripes on their backs. Some species have only one stripe. Sometimes people imagine these stripes are violin-shaped and think the spider is the extremely poisonous violin spider. The violin design on them, however, is very definite and would not be confused with the wolf spider.

Food of the wolf spider is mostly small insects like flies and mosquitoes. They are caught by pursuit for there is no web constructed. The wolf spider is often the prey of various wasps, and the battle between the two is very interesting.

The female wolf spider lays her eggs in a silken sack and then carries it around behind her like a caboose. When the young spiders hatch they normally cling to the mother and ride around with her. When she makes a kill, they get off and help her eat it. As they grow and become too large to all ride piggyback, they go off on their own.

Around our house we protect them. If I have to move them, I pick them up in my hand and put them outside. Anything that eats flies and mosquitoes is a friend of mine. How about you?

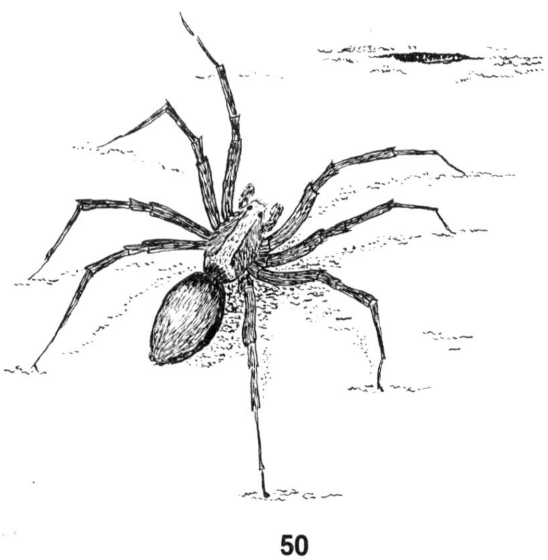

Wood Tick

It would be nice if I could tell everyone to go out and enjoy nature and not worry about anything. Such is not the case, however, because there are a few harmful things that everyone should be familiar with.

One of the least conspicuous of our not so friendly neighbors is the wood tick. These tiny relatives of the spiders have eight legs and a very flat brown body less than one-fourth of an inch long. They are known to take blood from a wide variety of warm-blooded animals of which man is one.

In order to feed, they bury their tiny head in the skin and then gorge themselves on blood. The males enlarge slightly while the females become greatly enlarged. All of this is done with little or no pain, and finding an enlarged tick on your body often is quite a surprise.

There are many procedures for removing ticks, some of which work and others that do not. The object is to remove the whole tick and not leave the head broken off in the skin. If kerosene is applied to the tick's body, it may make him easier to remove. A pair of flat tweezers can be used to grasp the body of the tick and then exert a straight steady pull for many seconds. If you pull too hard, you will pull the body off and leave the head. This will probably cause an infection and mean a trip to the doctor. If you can see the extremely small head when the tick is pulled out, then you will know you have been successful. Wash the area with soap and water or apply an antiseptic or both.

You probably will have no problems, but ticks can carry certain diseases like Rocky Mountain fever, tularemia, tick fever, and tick paralysis. Ticks get on grasses and brush and drop on their victims when the plant is brushed.

The ticks feed in each of their growth stages and in this way may become infected and pass along their infection to their next victims. Staying on paths, wearing clothes of tight weave and keeping them tucked in, and examining your clothes and body twice a day are good ways to prevent tick bites. An insect repellent that contains diethyl toluamide also is effective.

4
Mammals

When one identifies with the bouncy little bunnies or scampering, chattering squirrels, anything that hurts them must be bad.

One really has to understand the complexities of nature to realize that the predator is as necessary to the rabbits and squirrels as the nuts and grass they eat.

Mexican Free-tailed Bat

Hot summer days and warm summer evenings encourage people in our valley to relax out of doors at day's end. It may be a barbeque, a refreshing drink, or backyard gossip with a neighbor that takes one out of doors for a pleasant enjoyment of our evening breezes.

Sometimes seen at these times is one of the most unusual mammals known—the bat. There probably is not any group of animals about which there are more superstitions and weird tales than these unique creatures. Tales of witches' brew, vampires, and Halloween are but a few of eerie ways that we think of bats.

They are, of course, the only mammals capable of true flight. We hear of flying squirrels, flying lemurs and the like, but they can only glide, not fly. Flight is accomplished by what would be our hand with a thin membrane stretched between four of the fingers, the body and hind leg. The thumb is a free claw on the forward edge of the wing and, of course, the other fingers are greatly elongated.

All of the bats in our area are active in the evening or at night. This may have come about to reduce competition with birds for the supply of insects. All of our bats are insectivorous and are sometimes called flittermice. They are classified in the order or group Chiroptera. Their small beady eyes are probably of little use to them, and a fascinating sonar or sound system has evolved. In flight, a series of supersonic sounds are emitted that bounce back to the bat's ears. The sounds are interpreted with great accuracy, and the whole process is called echolocation. Another widely-known habit of our bats is hanging upside down by their feet when at rest.

There are several kinds of bats in our valley, but one of the most common is the Mexican free-tailed bat, *Tadarida brasiliensis.* This bat has a four-inch body and eight-inch wingspan. About half the tail projects beyond the membrane between the legs. This velvety-brown bat prefers attics and walls for roosting in our area but is one of the cave bats that roost in thousands in other areas. The musty odor they give off makes them unpopular when they roost around people. They generally migrate south in the winter rather than hibernate. Watch for them in the evening.

Bobcat

I wonder how many people have ever had the opportunity to enjoy the sight of a bobcat, *Lynx rufus.* Although this animal often lives on the outskirts of cities and is frequently active during the day, few people ever see him.

The bobcat or wildcat, as it is sometimes called, normally weighs up to 25 pounds. There are larger specimens on record however. They are catlike in appearance, and their fur is soft and deep. The general color of the fur in summer is light reddish-brown and fades to a gray in winter. The undersides and the inner surfaces of the legs are white spotted with black. The bobbed tail is four to six inches long with a spot of black on the top tip. The underside of the tip of the tail is white. This is one of the differences between the bobcat and its northern cousin the Canada lynx, *Lynx canadensis.* The entire tip of the tail of the Canada lynx is black. The Canada lynx is also larger, has bigger feet, longer tufts of hair on the ear tips and no spots on the undersides. The Canada lynx is not normally found in California.

Contrary to what many people believe, the bobcat is not much of a threat to birds. Numerous studies have shown that the bobcat's main diet is rabbits, ground squirrels, woodrats, and pocket gophers. Some birds are taken, however, and the bobcat readily climbs trees to hunt. The bobcat is also a good swimmer.

Typical of the cats, the bobcat does not dig its own den. Instead, it uses rock crevices, dense brush, hollow logs, etc. Mating normally takes place in the spring. There is even more yowling and screaming than a house cat. It is really a racket.

If captured young enough, bobcat kittens can be tamed but usually seem to be one-man animals. I will stick to house cats, however, since a bobcat kitten bit clear through my finger and would not let go. With a tetanus shot and the solicitous care of my doctor, all ended well. I am much wiser now.

Coyote

The role that a predator plays in our ecological system is often the most difficult for the viewer through nature's window to understand. When one identifies with the bouncy little bunnies or scampering, chattering squirrels, anything that hurts them must be bad. One really has to understand the complexities of nature to realize that the predator is as necessary to the rabbits and squirrels as the nuts and grass they eat. If the full reproductive potential of most of the plant consumers like rabbits and squirrels was realized, they would literally destroy themselves. They would consume all of their food supply and keep the food plants from reproducing. They could destroy their environment or at least reduce the carrying capacity significantly. In nature's intricate scheme the predator is very important.

Such is the role of the coyote, *Canis latrans.* Although he will eat berries or nuts and even carrion, he is primarily a predator. His 45 mile-per-hour speed makes it possible to run down jackrabbits while his patience and digging enable him to catch ground squirrels.

The coyote is about the size and form of a large collie dog. They run with their tail down while the ears are always erect. The long, grayish-brown fur makes up the basic color although there is a blackish stripe along the back. The belly is whitish while the nose, ears, back and legs may be reddish.

The high pitched yipelike bark is a real touch of wilderness to the nature lover. The coyote's clever ways often make it possible for him to live close to man. One was sighted in our creek area at American River College—a place surrounded by streets and houses. The uninformed sometimes get panicky when a coyote is near. They are far more timid than dogs, however, and are no threat to people. Although they sometimes help themselves to unattended chickens, lambs, etc., their destruction of pests far outweigh their damage.

I have heard it said that the coyote is the most obnoxious of all predators. If one is informed, however, he knows there are no "good guys" or "bad guys" in nature's world, only actors playing the roles nature gave them.

Mule Deer

In the fall the deer of California always receive a lot of attention. When the hunting season begins in the Sierra Nevada hunters take to the mountains in large numbers. Their quarry is the mule deer, *Odocoileus hemionus.* Mule deer, in the past, were often considered as several distinct species. Now, however, they are all classified as mule deer and six kinds are recognized. They are: Columbian black-tailed deer, California mule deer, Rocky Mountain mule deer, southern mule deer, Inyo mule deer, and burro mule deer. Reports of white-tailed deer are occasionally received from the northern and southern extremities of the state.

It is interesting that deer today are generally considered a mountain species. This was not always so. Historical records seem to indicate that in pioneer times deer were not as common in the high mountains as they are now, but they were very numerous in the valleys and foothills.

When the full impact of the gold rush caused settlers to crowd in, the deer population began to decline. Market and private hunting for meat and hides undoubtedly played their part, but the free grazing policies on public lands during the late part of the last century caused tremendous environmental damage and must be considered a major factor in the deer population decline.

There were so many grazing animals driven into the mountains in 1877 that W. A. Clarke wrote that because of a severe drought ". . . many flocks starved, and dying, were abandoned by their helpless owners. Some herders walked out of the mountains with their own blankets, their riding and pack stock being reduced to skeletons." Undoubtedly deer suffered greatly as a result of this damage. These factors along with several severe winters caused the deer population to reach a low ebb at the turn of the century.

Today, the deer herd in California is probably larger than it ever was in primitive times. Deer are the most abundant and popular big game animals in the state.

Mule Deer

Historically, the number of deer in California have fluctuated widely. At the turn of the century their numbers were at their lowest ebb.

Today, however, the numbers are probably much higher than in primitive times and certainly greater than the early 1900s.

Their increase in numbers is somewhat paradoxical since it has been concurrent with the increase in human population. It would seem that with the spread of human settlements, and the increase of soil cultivation, mining, logging, livestock grazing, and many other associated activities that changed the California landscape, that the number of deer would have been adversely affected as many other species have been.

Deer, like the coyote, however, often are encouraged by various of man's activities. When man cleared foothill and mountain homesteads, cut timber, and started forest fires he created much more open land.

This encouraged many palatable herbs and shrubs so that the deer habitat was actually improved.

Laws regulating the taking of deer were passed and were enforced by game wardens.

The ability of deer to live close to man along with their great adaptability allowed their high reproductive potential to be realized.

The fact that there are numerous deer is well documented by the damage to farm crops. They are often serious pests to apple, almond, grape, lettuce, bean, orange, avacado and alfalfa growers.

They also have done considerable damage to some of their own ranges. They have in some areas increased beyond the carrying capacity of their food supply.

They are literally eating themselves out of house and home. Malnutrition and die-off often follow.

It must be remembered that our deer herd is not completely natural anymore. We must realize this and recognize the need for a deer management program.

We owe it to the deer.

Reindeer

There is one popular animal that we all have heard about even though it only visits our valley once a year. Dancer, Prancer, Donner and Blitzen and all the rest of Santa's reindeer just come in December but other reindeer work all year long in other parts of the world.

The reindeer are members of the deer family, and among experts there is disagreement as to just what their species status is. Many scientists place all the reindeer and caribou in a single species, *Rangifer tarandus.* It seems that what we call caribou in North America is called reindeer in Europe and Asia. They are the only members of the deer family on which both male and female have antlers. Adults are more than three and one-half feet high at the shoulder and may weigh 600 pounds. There is great diversity in the shape of the antlers, but they are usually long sweeping beams with forwardly-projecting brow tines.

The feet or hooves of these animals are broad, flat, and deeply cleft. They are well adapted for walking on snow. Their feet make a clicking noise when they walk. The noise is evidently caused by a tendon slipping over a bone in the foot.

Coloration varies widely but is generally brownish or grayish. The underparts, inner legs and rump, are very light or white. The fur consists of not only the stiff, rather brittle and tubular guard hairs like our deer and elk have, but also has a heavy coat of woolly underfur.

Both the reindeer and caribou are gregarious although they periodically segregate by sexes. They are spirited and exhibit both curiosity and shyness. They feed on grasses, sedges, willows, birch, mosses and lichens (reindeer moss). A calf is able to follow its mother on the day it is born and must keep up as the group moves about.

The reindeer of Santa Claus is the type that domesticates best and is from Siberia. The same type is also found along the coast of Greenland and has been introduced into parts of Alaska and Canada.

To the people of the North, the caribou or reindeer is of vital importance for food and clothing. They are also quite important to many, many children in all parts of the world because of their magical flight.

Gray Fox

Most people are familiar with the crafty fox. We see one chasing a roadrunner and others stealing chickens in cartoons. Foxes are always portrayed as cunning, and that is an accurate description of them.

There are three kinds of foxes in California: the red fox, *Vulpes vulva,* the kit fox, *Vulpes macrotis,* and the gray fox, *Urocyan cinerocogenteus.* By far, the most common in California is the gray fox. This small collie-sized member of the dog family has a salt and pepper colored steel-gray back and sides. It has varying amounts of yellowish-red fur on the sides of the neck and on the legs. Black-tipped guard hairs make a dark line from the shoulders clear to the tip of the tail. The tip of the tail is black, while the tip of the tail of the red fox is white. The kit fox and coyote have the black only on the tip of the tail. The gray fox has a black face and muzzle, while the back of the ears are reddish-brown. The belly and chin are white.

The western slope of the Sierra is the primary area for the gray fox. He is found in the yellow pinebelt and the chaparral and wooded areas of the foothills and valley. You may see him when he crosses roads, climbs rocky areas, or scampers under shrubbery during the night or day. Hunting is done when food is available, and most of the food is animal in nature. Pocket gophers, mice, other rodents, rabbits, insects, some birds and eggs make up most of the food. Plant food includes berries, grapes, and some other domestic fruits.

Rarely digging his own den, the gray fox prefers crevices in rocks and hollows in logs. He can climb slightly leaning trees almost as nimbly as a cat and is the only fox that climbs trees.

The wiry texture of the gray fox fur kept it from being fashionable in the United States. Nevertheless, it was important at one time because of the great numbers taken.

The animal is protected as a furbearer and is, in general, a valuable predator. You will be lucky if you ever see him.

Pocket Gopher

The valley pocket gopher, *Thormomys bottae,* is the only species of gopher in our valley. The head and body size is variable as is its color. The head and body may be almost five to seven inches long with a tail from two to almost four inches long. Although their color may vary from almost white to black, most are shades of brown.

The pocket gopher is a fossorial or burrowing animal that is well adapted to a life of tunnelling in the earth. The forelimbs are very sturdy and the spatulate hands have huge claws. The eyes and ears are small and the external cheek pouches or pockets are fur-lined and reversible. Their large yellowish incisor teeth are always exposed in front of the mouth opening even when the mouth is closed. In a unique arrangement the skin with hair extends behind the incisors. The teeth are often used to cut roots and as picks in digging in hard soil.

Gophers live on green vegetation, bulbs and roots. Although they do not usually venture out of their tunnels, they may occasionally surface at night. Normally, they burrow just below the surface to get the roots and bulbs and will pull green plants down into their burrows from below.

The gopher digs by clawing and biting the soil and then pushing it under and back with his paws. When there is enough to plug the hole, the gopher turns around and pushes the dirt up and out of the hole and leaves it in a fanlike arrangement. When finished in that part of his tunnel, he pushes the last of the soil up and plugs the entrance. This process serves as an effective soil mixer and also loosens the soil for water penetration.

The pocket gopher is a valuable ally in the wild but an unwelcome guest in yards and gardens. At least in mine he is.

Mole

There are several kinds of animals that people may confuse with one another. Some typical examples are: rats with mice, chipmunks with ground squirrels, and moles with gophers. Although the relationship between rats and mice and between chipmunks and ground squirrels is very close, the gopher and mole are quite distantly related.

Gophers are rodents with well-developed incisors or front-gnawing teeth. Their main food is plant material. Moles are insectivores—a different major classification of mammals. The insectivores have many sharp-pointed teeth and feed on a variety of other animals.

The mole in our valley is the California mole, *Scapanus latimanus,* which has a stocky body about 4 1/2 inches long with a 1 1/2 inch slightly-hairy tail. The cylindrical body is covered with a short, thick, velvety fur that can lie in either direction. This enables him to move either direction in his tunnels. The fur is a blackish-brown to black and covers the eyes and ears so they are all but invisible. The most obvious characteristic, however, is the rather useless-looking front feet. They are actually broader than they are long. Because of this, the California mole is often called the broad-footed mole. He uses these feet like blades of a digging machine and is able to dig with amazing speed.

Evidence of this mammal are "mole runs" or long ridges of raised earth on the surface of the ground. These are made as the mole forces and digs his way just below the sod level looking for worms and insects. They are easily told from gopher's workings because the mole raises the sod and all. The traces a gopher leaves in the mountains are ridges of bare soil that have been carried and pushed up into the snow. When the snow melts, the plugged tunnels remain as ridges. Moles often leave their mole hills which are conical-shaped piles of bare earth while gophers leave mounds that are fan shaped.

House Mouse

The changing of the seasons brings many surprises to those who watch through nature's window. Many of the surprises are pleasant while others are rather perplexing. When the rains come in winter they awaken many of the resting plant seeds, coat the mountain tops with icing, make soil spongy under our feet and cause many of nature's smaller creatures to search for more suitable quarters. One of these creatures that often moves in to our houses with us is the house mouse, *Mus musculus.* Every winter I think our home is mouse proof, but once or twice I find their small one-quarter inch spindle-shaped droppings and out come the traps.

The familiar house mouse is not a native but probably originated in Asia. It has emigrated all over the world by boat, trains, trucks, etc. It is usually thought of as a resident of buildings, but in our valley it can survive very well in grasslands often a mile or more from dwellings. The rather uniform brownish-gray back, buffy or white underparts and long sparsely-haired tail is familiar to all of us. The 3-4 inch long body that is almost doubled by a 3 1/2 inch tail weighs only about 1/2 ounce.

It is difficult to realize that a mammal this small can have such great reproductive potential. Young can appear at any time in litters of 4 to 11. They are born 21 days after mating, and the young can mate at 2 months. Theoretically one pair could have 1,000 descendants in one year if there was enough food, no enemies or disease. With a life span of four years just imagine what the potential of one pair is. In California one acre was known to contain 82,000 individuals.

The hiding or storage tendency is poorly developed in the house mouse, and they do not hibernate. Cleanliness is not practiced, and anytime they are concentrated the musky smell of their urine is very evident. Next to the brown rat, the house mouse is probably the most destructive mammal there is. It can carry diseases, injure woodwork, eat grain, etc.

The house mouse is also the white mouse of the pet store and laboratory and is used extensively in research. The many color variations make this mouse very popular. Next time you are in a pet store take a look. I am sure you will agree.

California Vole

There are many animals that most people do not get a chance to see well. The larger ones like deer, raccoons, and squirrels will come in contact with people at some time, but some of the smaller ones that stay in the wild might never be seen at all. One of these unfamiliar animals is the vole or meadow mouse.

In our valley we have the California vole that is also known as the California meadow mouse, *Microtus Californicus.* Most of the voles look much alike with long fur, small ears, black beadlike eyes, and a relatively short tail. These characteristics give the voles a much different appearance than the house mouse or white-footed mouse. These mice both have larger eyes and ears, shorter fur, and a much longer tail.

The California vole has a head and body that may be almost six inches long and may have a tail almost three inches long. Their weight will sometimes run over three ounces. The overall color is a grayish-brown with pale-colored feet and under tail. They like marshy ground, wet meadows and dry grassy hillsides. They are active both day and night and are what hawks, owls, snakes, cats and other predators often hunt for and feed on.

Their presence is often detected by runways in the grass one to two inches wide. There are often small piles of droppings and short pieces of grass along the runways as a further indication.

The reproductive rate of these animals is phenomenal. One female may produce 13 litters a year which may be about 100 mice. The female may breed when one month old and the young are born 21 days later. The female breeds immediately.

If you would like a surprise, take a pair of mice and see how many offspring that pair and all their litters could produce in a year.

Muskrat

If viewers through nature's window spend much time near water in our valley, they will come across a rabbit-sized rodent called the muskrat, *Ondatra zibethicus.* This plump, robust, little fellow has dense, glossy fur, small ears, a nearly hairless, scaly tail that is flattened from side to side. His hind feet are partially webbed. His back and sides are usually a rich brown color with the sides being lighter. The underparts are pale with the throat being almost white, and there is often a dark chin spot. The actual length of his body is 16 to 25 inches with a 7 to 11 inch tail. Weight may run from 1 1/2 to 4 pounds.

Muskrats are found in both salt and freshwater marshes and along the edges of most waterways. They may travel overland for several miles, and many are killed on highways considerable distance from water. I have also been called upon to rescue them from swimming pools.

Their diet consists of the stems and fleshy parts of many plants like cattails, rushes and various grasses. They will also eat freshwater mussels and certain other invertebrates.

The muskrat gets its name from two small glands on the lower abdomen that secrete a musky substance during the mating season. This scent is used to attract the opposite sex. Before perfumes were available, ladies would place their kerchiefs over unskinned muskrats till they absorbed the fragrance for the same purpose.

The muskrat was native on the eastern side of the Sierra Nevada but was not found in other parts of California until it was introduced. In 1920 the eastern muskrat was introduced into northern California and in 1938 into the Feather River drainage. From there it spread throughout our valley.

The soft fur is thick and durable, and its versatility makes it the "backbone" of the fur industry. They are often sold in the fur trade as Hudson Seal, Russian Otter, Red Seal or River Mink. Its durability is 45 percent that of otter.

In irrigation areas the muskrat can create many problems with his burrowing. In southern California they are not protected, but in the northern part of the state they can be taken only during the trapping season. This little fellow is now an established member of our valley.

Opossum

One of the most unique mammals in California is the introduced Opossum, *Didelphis marsupialis.* This cat-sized ratlike mammal is a member of the order Marsupialia or pouched mammal. This is the group that is dominant in Australia. The opossum is the only marsupial found in North America and was formerly restricted to the eastern states. In 1906 some were introduced near Los Angeles and in 1912 near San Jose. They established themselves very well in the warm cultivated lowlands and along well-protected waterways into the mountains. They are now very common in much of California and often wander into garages, under houses, and in front of cars.

When they are discovered, they may play "possum". Their tongue will loll out on the ground, their eyes are half closed, and their jaws may drool. They may feign death even to the slowdown of breathing and heart rate. When the danger has passed, they may quickly revive and creep away. Researchers are not unanimous in their explanation of "playing possum" and do not agree on its cause. Some say it is a true "faint" while others profess it is just pretend.

I will have to admit that I do not qualify as a opossum scarer as I have never seen one "play possum" even when I tried to scare one.

The opossum is essentially nocturnal and eats fruits, vegetables, nuts, meat, eggs, insects and carrion. The tail is prehensile and ratlike while the feet are well adapted for climbing.

The young are born very immature and must travel to the pouch unassisted. They are so small a dozen would fit in a teaspoon. In the pouch they remain fastened to a nipple. In a month they are about the size of mice. In three months they are too large for the pouch and may travel hanging onto mother's fur. Sometimes they may even use her tail for additional support although I have never seen them do this.

Watch for this unique intruder into our valley. He is definitely one of our most interesting animals.

River Otter

One of the most fascinating animals in the world is seldom seen by the casual nature enthusiast. One would think that since this mammal is not seen often, that he is very shy and retiring. Such is not the case with the river otter, *Lutra canadenses.* This animal is quite active and adventuresome in his generally remote habitat. His humorous behavior has earned him the nickname clown of the animal world.

The adult river otter is about four feet long including the rather long full tail. He may weigh from 10 to 25 pounds. Their luxurient brown fur is quite dark when wet but is considerably lighter when dry.

Young otters are called pups, and a normal litter size is two or three. Their den may be a burrow in a bank, a hollow tree, or something similar. The mother and young usually stay together till the next litter is due.

The frolicsome antics of the flamboyant otters are often due to their method of chasing fish down in shallow streams. Fish serve as the main food but many other organisms like clams and crayfish are consumed. In our area, dead and dying waterfowl are also eaten.

This member of the weasel family is probably the most adept swimmer of all nonmarine mammals. All four feet are webbed and his body is beautifully streamlined. The dense fur is excellent insulation against the cold water.

The adult otter has practically no natural predators. The main threat to their survival, however, is man. It is illegal to kill, capture or possess a native otter, but many are caught and killed accidently in steel-jawed traps set for muskrats—a very sad end for any animal.

Raccoon

One of the most familiar ecological partners we have is seldom seen by those who look through nature's window. This sounds contradictory but nevertheless is very true. I am referring to a mischievous and pleasant creature of the night—the raccoon, *Procyon lotor.*

This lovable little character may weigh up to 25 lbs. and have a head to tail length of 30 inches. The black and gray ringed tail may be another 10 inches. The black face mask, ringed tail, and food-washing habits are almost as well known as the family dog and cat.

The range of the raccoon is interesting since it is found over most of the United States except the northern Rockies and the drier parts of the Great Basin. Well watered wooded areas provide him with his necessities. The very agile front paws work like our own hands and make catching frogs, crayfish, insects and snails easy. Various fruits like grapes and melons are also much sought after, and in some areas his corn-eating habits get him in a lot of trouble.

His highly inquisitive nature and playful antics seem to catch the fancy of old and young alike. When obtained young, they make fine pets. As they get older, however, they have quite a mind of their own and can get into a great deal of mischief.

I had a raccoon at American River College for eight years, and he was always very friendly. When very young he was much like a playful puppy and would follow us around everywhere. As he grew older however, he became very strong and went where he chose and had to be coaxed with tidbits of food.

The raccoon does not hibernate although various authors seem to disagree. It is fascinating to see them start storing fat during the summer till they literally waddle around in the fall. This stored energy enables them to get by on lean rations during the winter.

Hunting raccoons at night has long been a favorite sport of some. Coons dogs would chase the coon till he climbed a tree. He would then be shot. Today the true sportsman trees the coon and then goes home. He leaves this fascinating and clever little fellow for another chase. This makes everyone happy, especially the raccoon.

Jackrabbit

There is only one kind of jackrabbit found in our valley—the black-tailed jackrabbit, *Lepus californicus.* Its long ears and hind legs make it easy to distinguish from our brush and cottontail rabbits.

Although it is a "jackrabbit", it is a hare not a rabbit. Rabbits dig a burrow and make a real nest for their young. Rabbit babies are born with their eyes closed, no hair, and need much mother care. Hares, however, smooth out a surface nest and give birth to young that are fully furred, open eyed and almost ready to run. They are probably nursed for only a few days.

The name jackrabbit was applied by early pioneers who likened the "jacks" ears to those of a jackass. They called them the jackass rabbit. Later this was shortened to a jackrabbit.

The upper body is a pale yellowish brown that is ticked with black. The undersides are buff to white. The back of the tips of the ears and the top of the tail are black. They are about cat sized with a spot of white usually present on the forehead.

Jackrabbits prefer open weedy areas, and as long as these are present will live close to man. I have seen upwards of forty sitting on the Haggin Oaks golf course fringe areas. When being chased they can run about 45 mph and leap 15 feet. One author says they can leap 25 feet. They are found all over our state except in mountainous areas over 12,000 feet.

They are active mostly at night, and the mother will hide her young when she goes out to feed. They are strictly herbivorous, eating various plants including the bark of trees. In large numbers they can do considerable damage, but they have many predators seeking them for food. Coyotes, foxes, hawks, eagles, and owls are but a few. Dogs, cats, and automobiles also take quite a toll. Although the jackrabbit doesn't have great beauty to recommend him, you have to appreciate him as a very successful environmental partner.

Rabbits

The rabbit's foot has long been known as a good luck symbol. I don't know how this superstition got started but maybe it arose because of the rabbit's skill in escape. The rabbit's foot that one would use for good luck did not, however, bring that particular rabbit any luck at all.

There are three so-called rabbits that could be seen in our valley. One is the jack-rabbit, which is a hare and not a rabbit.

The two true rabbits we have are the cottontail, *Sylvilagus audubonii,* and the brush rabbit, *Sylvilagus bachmani.*

The brush rabbit is smaller than the cottontail and is about 12 inches long. His ears are shorter than the head, and the body appears dark brown with heavy washes of black. They are seldon seen out of the heavy brush they inhabit.

The yellowish-brown body of the cottontail has a blackish overwash. The ears are about the length of the head, and his length is about 15 inches. They are common where grassy areas are fringed by shrubs and the thickets that provide good escape cover. They are not seen too often because of their early morning and evening feedings habits. They seldom venture more than 25 feet from escape cover and generally disappear at the first sign of danger.

The reproductive capacity of cottontails is well-known. From February through September a female may come in heat every week until she is mated. Four or five litters a year are possible with three to six in each one. Each young weighs about one ounce and is hairless, blind, and completely helpless. Nevertheless, in four weeks they are weaned and driven from the nest.

The survival of the cottontail depends on elusiveness and not speed. Their top running speed is only 18 to 20 mph, so virtually any dog can run down a cottontail in open country. The rabbit does have an edge in awareness, however, and hearing is their most developed sense. Sight is next, and since the circle of vision is almost complete, it is almost impossible to "sneak" up on a rabbit. Try it and you will find out.

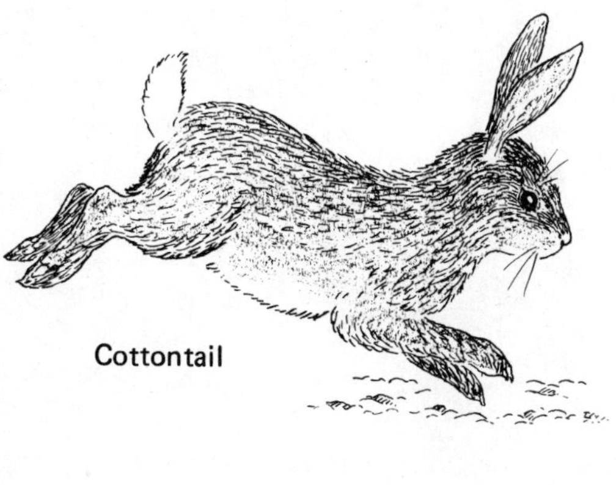

Cottontail

Brush Rabbit

Easter Bunny

He hops up and down, he hops all around,
He hides eggs in bushes and also on the ground.
The kids try to see him but never get a chance,
But when they find the eggs, they do quite a dance.

I'm sure the kids all know who this rhyme is all about,
It concerns the Easter Bunny there is no doubt.
So watch for his nose and big floppy ears,
And remember he's been coming for many, many years.

The Easter Bunny is a very special type of creature and it is quite difficult to find accurate scientific information about him. I will tell what I have found, however, and you will just have to take my word for its authenticity.

The size of the Easter Bunny, *Bunnius Easterensis,* is largely a result of learned imaginaton. For some, he is man-sized while for others, he is rabbit size. Color is also quite variable, probably affected by tradition and creative need. In other words, he can be almost any color.

We know little more about the habitat of the Easter Bunny than we know about his size and color. He evidently stays out of sight completely, and the only signs he leaves are the colored eggs at Easter time. He seems to be particularly attracted to houses where children live. He usually hides colored eggs, candy, and sometimes gifts for deserving children when he visits at Easter. Nobody seems to know where the Easter Bunny gets the gifts he brings, but then I suppose the only people who worry about that are parents.

Skunks

Occasionally in the world of nature there are creatures better known by some characteristic other than their appearance. Such is the case of the skunk. Almost everyone is familiar with the malodorous spray of the skunk, yet few actually have seen one alive.

There are two kinds of skunks found in California, the striped and the spotted skunk. The striped skunk, *Mephites mephites,* is about the size of a house cat. It is easily recognized by its black body and narrow white stripe from the forehead that splits into two stripes down the back to the base of the tail. The tail may or may not have a white tip.

The spotted skunk, *Spilogale putorius,* is about the size of a medium squirrel. Its black body is characterized by a white spot on the forehead and under each ear. There are usually four broken white stripes along the neck, sides and back.

Both species are nocturnal and belong to the weasel family. They eat a variety of food like fruits, berries, eggs, insects of all kinds, small rodents and reptiles.

If possible, they usually dig underground burrows or will live in haystacks, under roots, rocks or old buildings. Skunks do not hibernate but may sleep during spells of cold winter weather. There are sometimes several found in one den.

Skunks are well-known for their ability to spray a very obnoxious sulfide liquid for defense. A normal release may be only a twentieth of a teaspoon or one drop. The odor may carry more than half a mile, however.

Skunk fur in the 1920s brought as much as five dollars a pelt. Coats were sold under the trade name Alaska Sable and Black Marten. Today a pelt is worth less than a dollar. Few skunks are trapped now for their pelts. By habit skunks are more valuable than destructive, but in Sacramento County many skunks have been found to be rabid. If a skunk is seen during the day, beware. It may well be rabid and the County Animal Control Center should be called.

Western Gray Squirrel

The antics of many animals intrigue most of us but none are more fun to watch than the squirrels. There are several kinds of squirrels in California like chipmunks, California and golden-mantled ground squirrel, pine or Douglas squirrel, and the gray squirrel.

The California or Beechy ground squirrel is the common squirrel of our valley, and many people think it is the only one. It is not, however, because the Western gray squirrel, *Sciuris griseus,* is often found along waterways where there are plenty of oaks.

We often see this slender, strong-muscled squirrel along the creek of our American River College campus.

These born acrobats are well adapted to an active life in the trees. They have gray bodies with white underparts, and very bushy tails. The body length is ten to eleven and one-half inches long with the tail being almost the same length. They weigh one and one-half to two pounds.

The principle food is pine seeds and acorns which are available over a long season. The pine cones are often cut from the tree by the squirrel. When they fall, the squirrel above can sometimes be seen. The squirrel then descends to the ground, chews the scales from the cone by starting at the base and eats the seeds. The piles of cone scales are obvious signs of the presence of squirrels. Acorns supplement the diet of pine nuts and are buried in the ground singly. They are found by smell later and dug up for food when needed. Some are not found, however, and many oaks are undoubtedly the result of the gray squirrel. Gray squirrels do not have cheek pouches and usually eat food where they find it.

Gray squirrels are common residents of the foothill and yellow pine belts of the Sierra Nevada from 400 to 6,000 feet. They do not hibernate but hide in their tree dens in bad weather. They are active throughout the year, and many move to lower elevations during the winter.

California Ground Squirrel

The California ground squirrel, *Otospermophilus beecheyi,* is probably seen by more Californians than any other native mammal. Who is not familiar with this little fellow sitting on a fence post, hunched near his hole in a field, or scampering back to his hole from a nearby tree?

His dull yellowish brown fur is ticked with black and white hair ends to give him a rather mottled appearance. The sides of his neck and shoulders are a grizzled white almost making it look like a collar. There is often a dark brown patch shaped like a cape over the shoulders and part way down the back.

Often called the digger or beechy ground squirrel, they were once enormously abundant. They are found throughout the state except for the highest mountains, the Colorado and Mohave Deserts, and east of the Sierra crest. They are the only common squirrel in our valley although the gray squirrel is sometimes seen along waterways and the eastern fox squirrel may be found in some of our parks and city streets. Neither of these have the field marks of the California ground squirrel and should not be confused with it.

As indicated by their name, they are ground dwellers and cause much trouble with their digging. They inadvertently cause a great loss of irrigation water by tunneling in levees, canals and river banks. They can be very destructive to agriculture by eating nuts, fruits, and vegetables. Their holes may be stepped in by farm animals causing broken legs. It is also known that their fleas can carry bubonic plague. All of these problems do not make the California ground squirrel popular, but it does make a good case for protecting predators like coyotes, hawks, foxes, and bobcats. Many farmers have now learned how beneficial predators can be in keeping the ground squirrel population down, and many now actively protect the predators.

It must be remembered, however, that there are other ecological aspects to the California ground squirrel. Their burrows may serve as shelter for toads, snakes, skunks, and the burrowing owl. When one looks closely through nature's window, he soon finds that each life plays a part in nature's scheme.

Chipmunk and Golden-mantled Ground Squirrel

Summertime to most Californians is a subtle signal for frequent mountain visits. Taking the children to a mountain campground is one of the most popular pastimes. If you doubt this statement, just check the mountain highways on a summer weekend.

One of the reasons for the popularity of our high country is the diverse and fascinating scenery. Another reason is the friendly wildlife present in most campgrounds. Children and adults alike are especially enchanted by the many squirrels. It is too bad that most parents do not know the difference between chipmunks and golden-mantled ground squirrels. They use the term chipmunk indiscriminately and, as such, actually give their children erroneous information. It is very easy to tell the two apart, and perhaps this article will help clear up many misconceptions.

Both of them are rodents, of course, and also are in the squirrel family. They both have cheek pouches inside their mouths that they stuff with food as they gather it. This food may be eaten, buried in shallow holes and then covered with dirt, or it may be taken to the burrow. Both live in some kind of an underground nest and hibernate for the winter.

The golden-mantled ground squirrel, *Citellus lateralis,* is larger than any of the chipmunks. They are about two- thirds the size of a house rat and are often called "copperheads" because of the yellowish or coppery red head and neck. The black-bordered white stripes that run along the side of the back do not cross the shoulder onto the face. These are often the tamest of our ground squirrels.

Chipmunks belong to the genus *Eutamias,* but there are eight species that may be found in the Sierra Nevada. They are smaller than the golden-mantled and have alternate dark and light stripes. There are nine of these stripes on the back and 5 on each cheek. There is also a white spot behind each ear.

Watch for these friendly fellows on your summer trips and look for the stripes on the face. If there are none, they are golden-mantled ground squirrels. If there are stripes, they are chipmunks. Very simple.

5
Birds

One of the joys of having an interest in nature is the intriguing way her secrets are unfolded.

When one first starts to observe, only the obvious is seen. As one becomes more attuned, an ever-increasing crescendo of delights is available.

Bird Names

When many people first start noticing the birds in our area there are a few species that they think they see but really do not.

The most common mistake made is referring to our local scrub jay, *Aphelocoma coerulescens,* as a blue jay. I have to admit that our bird is a jay and it is blue, but its name is not blue jay. The Steller's jay, *Cyanocitta stelleri,* found in both the coast range and the Sierra Nevada is also quite often called a blue jay. It is a jay and it is blue, but his name is still not blue jay.

The true blue jay, *Cyanocitta cristata,* is the common jay of the eastern United States. He is a large bright blue bird with a blue crest. There are white spots in the wings and tail and a black band or necklace across the throat. This very sassy eastern bird has a dubious reputation, but there is no question that he really livens up the area when he moves in. His habits and character are very much like our own jays.

Another common mistake made is referring to several of our woodpeckers as red-headed woodpeckers. Many of them do have some red on their heads such as the acorn, the male Nuttall's, hairy, downy, and the various sapsuckers. Only our red-breasted sapsucker, *Sphyropicus varius,* really has a completely red head. The true red-headed woodpecker, *Melanerpes erythrocephalus,* is not found in California at all but occurs mainly east of the Rockies. This black-backed bird has a white rump and large square white patches on the rear edge of the wing. They are very conspicuous in flight. The entire head is a bright red.

It is understandable why so many people call our local birds by these incorrect names. We see them on calendars, playing cards, posters, etc., so often many people think they are the only kinds of jays and woodpeckers there are. I hope teachers and parents alike will use the correct names so children will learn our birds here. If you don't know which kind of jay or woodpecker it is, just call it a jay or a woodpecker. Then you are correct.

Bird Songs

The sounds of spring are many and varied. They are produced by many types of nature's creatures and undoubtedly express many different needs. Bird song is one of the most frequently identified and probably the least understood sound in nature.

Most of the observers through nature's window assume a bird sings out of joy or happiness. It is a logical conclusion since we frequently do just that.

Those that have studied bird songs, however, tell us that their singing is for very different reasons. Most birds sing only during the mating and nesting season, and are relatively silent the rest of the year. In fact, it is probably not accurate to call the sounds a bird makes singing at all. It is more accurate to refer to his vocalizations as a call.

One of the primary calls a bird makes is to establish his territory. It is this call that most people notice and assume he is singing his heart out because he is so happy. He is, instead, telling other males that this territory is taken and intruders beware. He "stakes" his claim by flying from one point to another and voicing his sentiments from each. It is these calls one hears from fence posts, TV antennas, and other prominent places.

Closely associated with the territorial call is the mating call. For several species the calls are the same and serve to attract a mate as well as establish a territory. Investigators

have isolated other different types of calls that indicate distress, alarm, food, danger, brood time and probably many others.

The next time you hear a bird vocalizing, remember he is really saying something. Can you tell what it is?

Injured Birds

The magic of springtime holds endless fascination for those who enjoy the wonders of nature. It is also, however, a time of great emotional trauma when one finds a baby

bird in the mouth of a cat or dog. I am asked quite often what to do for these helpless little creatures and I have some general instructions I pass on.

The best advice I can give is to try to put the baby bird back in the nest if it is not injured. If the bird can fly a little, perhaps just putting it in the tree is good enough. Do make sure it is a tree-nesting bird, however, and not a killdeer. Often, the parents will feed the young birds wherever he is—even in a cage sometimes. The most common baby birds I come in contact with are scrub jays, robins, mockingbirds, yellow-billed magpies, starlings, Brewer's blackbirds and house sparrows. They all are common yard nesters.

If the nest of the young bird cannot be found and you decide to raise the bird yourself, you will find out just how hard parent birds work in the care of their young. Although there are several kinds of food that can be used, I have had the best results using a basis of dry dog food softened with water to a mayonnaiselike consistency. At first, the young bird's mouth may have to be forced open, but it usually isn't long till the mouth will open automatically and demand food every few minutes. The dog food can be supplemented by chopped egg yolk, worms or insects as often as possible. When the bird is grown and can fly, it must be released as it is illegal to keep any but the starling or house sparrow. The local State Game Warden should be notified when you are keeping any bird but the two I mentioned. This is especially true for any hawks and owls.

If the young bird is injured, the mortality rate is much greater. Sometimes veterinarians can help, but it may be expensive and disappointing.

Bird Feeders

With the end of summer those who enjoy the view through nature's window should be getting ready for our winter visiting birds. It can be a great satisfaction to attract these friendly fellows into your yard. A well-supplied feeding station should do the trick.

A feeding station is any place that bird food is put out regularly. It could be a place on the patio, a platform in a tree or post, a piece of plank on your barbecue, or any of several other devices. Birds are far more interested in the food than how fancy the feeder.

Any feeding station should be placed so that cats cannot hide too near. The area around the station for about five feet should be open, and escape cover should be available from five to ten feet away.

Many types of food can be used. Bread crumbs are good but blow away. Wild bird seed can be purchased at feed, pet, and grocery stores for a nominal price. It, too, may blow away however. Suet, the hard fat from beef, can be put out in chunks, smashed into the bark of trees or put in holes of logs. Unsalted sunflower seeds, raisins, nuts, uncooked fruits and cereal, and a variety of other things may also be used.

My favorite food for birds takes a little effort to prepare but is very convenient. Render some suet over a low fire. While still warm, mix with a mixture of many odds and ends found in the house. I use corn meal as a base and like to put in bird seed, sunflower seeds, raisins, old cereal, cracked walnuts in the shells, suet chunks and peanut butter. I cut the top out of a half-gallon milk carton and fill it two-thirds full of my mix. I then pour in about a cup, maybe more, of warm liquid grease and stir until the grease is well mixed with the meal. Settle the mix by hitting the bottom of the carton on the table. A hole can be punched for later hanging if desired. Let the mix harden overnight. Extras can be put in the freezer for next year even. The whole block can be put out or cut into pieces. Rain won't hurt it, and it will not blow away. It will attract a large variety of birds. In fact, even your dog may like it.

Western Meadowlark

Perhaps the most typical bird of the West is the western meadowlark, *Sturnella neglecta*. Not only is this bird common to the broad expanses of grasslands most people associate with the West, but it is one of our most common birds. This member of the blackbird family is not, as its name would indicate, a lark. It is well noted for its song, however, and its clear, flutelike trills put the eastern meadowlark to shame. The eastern, *Sturnella magna,* and western meadowlark are very similar in appearance, but we have only the western here.

The upper parts of the western meadowlark are brownish with black streaks and a buffy line through the crown. There is a yellowish line from bill to eye that continues as a buffy line behind the eye. The underparts are a beautiful bright yellow with a striking black crescent across the breast. The extremely short tail has white outer feathers that make a good field mark when the bird flies its typical short glidelike flight.

Meadowlarks hide their nests very well in tall grasses. They are made of dry grasses fashioned into a compact cup that is lined with finer materials. A dome-shaped roof is present on most so that the nest and its occupants are hidden from the searching eyes of crows, hawks, jays, skunks, weasels, and snakes. A well-worn path usually winds its way through the grass to the nest so that the nest is not approached directly. Normally, two broods of five to seven are raised.

The food of the meadowlark is over 70 percent animal matter and as a destroyer of grassland pests like cutworms, caterpillars, and grasshoppers, is probably unequaled. Undoubtedly some seeds and grain are also eaten.

Large flocks of western meadowlarks form in the fall and winter. They gather in such large numbers in our area that the Sacramento Audubon Christmas Count typically counts more of them than anywhere else in the nation.

It would be a sad day indeed if our environment were so altered that the inspiring song of the western meadowlark disappeared. Listen in the spring and see if you do not feel better for having heard his song.

Red-winged Blackbird

There are few birds as well-known as the red-winged blackbird, *Agelaius phoeniceus.* When we drive along roadsides in the very early spring it is not at all unusual to see male redwings spaced rather evenly along the strands of barbed wire. They are also found near patches of tules or cattails of almost any size, and near lakes, streams, and marshy areas. The brilliant red shoulder patches of the coal-black male are known by all but the brown-striped female is much less noticeable.

The male selects the nesting site, and with wings spread and fluttering shows off the brilliant red patches to attract a female. When a female accepts the male, she accepts the nesting site. The nest is built of plant materials like grasses and is woven beautifully around cattails or tules. It is basket-shaped and built well above the water level. Three to five one-inch long eggs that are bluish and streaked with black are laid. Only the female incubates the eggs and for a period of 10 to 14 days. It is possible a pair may raise two broods in a season.

While the female is incubating the eggs the male puts on quite a display. Using his wing color as part of a threat display, he establishes his territory by flying to the outer edges of what he considers his and sings loudly. He darts out at any and all intruders with great bravery while the female sits

quietly on the nest. He will even fly out and attack a hawk if the hawk comes too close.

Most of the food of the redwings is insects, and a flock can serve as an effective control on certain species. At some times, however, large flocks may cause damage to agricultural crops. Their overall beneficial effect is well recognized.

The males and females go their separate ways after the young are raised. The males flock together, and the females stay in their flocks. I have seen the flocks mixed in winter but they show no interest in one another.

Brewer's Blackbird

When watchers through nature's window see huge flocks of robin-sized blackish birds, they will probably be seeing starlings or Brewer's blackbirds or both.

The Brewer's blackbird, *Euphagus cyanocephalus,* is the commonest and widest ranging of our blackbirds. In fall and winter they may gather in flocks of thousands to feed and roost.

The male is essentially black, but in late

winter he becomes very glossy, and in strong light purplish, greenish and blue hues are evident. The iris of his eye is a striking white. The female is more dull and the head appears brownish or grayish while her body is also black. She can easily be told from female red-winged blackbirds because she is not streaked. Her iris is dull brown.

Brewer's blackbirds have a very pleasing birdlike appearance and are much more attractive than the starling. They have a longer tail than the starling and a sharper, more slender beak than the brown-headed cowbird. However, both the starling and the brown-headed cowbird can be confused with blackbirds.

The Brewer's blackbird is a western species found west of the eastern rise of the Rockies. They like to nest in low bushes and trees. The female builds a rather coarse nest of grass and twigs formed with mud. Those nesting on American River College campus are very sassy nest defenders and may fly at you or actually peck you as you walk by. Two broods may be raised in a season.

Without investigation, the feeding habits of this bird would seem to doom it to unpopularity. It eats fruit and grain and feeds in stock pens. It also eats, however, weed seeds and insects. They will often desert fruit or grain fields to feed on freshly plowed ground. Except in certain instances, the Brewer's blackbird is usually beneficial to us and makes a very handsome ecological partner.

Starling

There are some things that are difficult to ignore when one looks through nature's window. Earthquakes, tornadoes, cloud bursts and lightning get everyone's attention whether they are looking or not. The starling, *Sturnus vulgaris,* is almost that obvious and it seems as if he is everywhere one looks. He is in the trees, on power lines, in the air and on the ground.

His black and glossy-green color are set

off by a yellow bill in the spring while in winter his coat is black with speckles of buff or white. His body is about the size of a blackbird but his tail is much shorter. If one is the least bit observant, pairs of starlings will be seen perched almost everywhere. Their rather upright posture is very characteristic and their plumage is so glossy it is almost unbelievable.

The starling is an introduced species in the United States and got its start before any regulations against importations existed. Although there could have been other releases, the information I have mentions 80 birds released on March 6, 1890 and 40 more April 25, 1891. Both releases were made in New York's Central Park. The poor misguided importer probably thought the starling would eat harmful insects here as it does in its native home in Europe. How mistaken he was! Instead of helping us with insect problems, they have caused tremendous fruit crop damage, eaten livestock feed, and driven out native birds.

After their introduction, they spread throughout the United States and reached California in the 1940s. Once here, they reproduced rapidly by eating livestock feed, grapes, olives, cherries and many other fruits. To nest, they may drive the acorn woodpeckers and other hole nesters from their own homes. They successfully forced their way into our natural scheme and can be seen everywhere. Watch for this handsome but undesirable intruder. He'll be watching you.

Starling

Mallard

Few people in our valley realize the tremendous influx of birds we receive in the fall and winter. It would seem that the drab lead-gray skies, chilling winds, and deluging rains are inhospitable to winter bird populations. Compared, however, to the north and high mountains from whence they came, our valley is a welcome refuge from the deep sterile snow that blankets their summer range. Actually, our climate is quite hospitable and comfortable for our visitors.

There are many species of birds that spend the winter with us, but none is as spectacular as the millions of waterfowl that utilize our meager wetlands. Of all the ducks and geese that visit us, none is more well-known than the mallard duck, *Anas platyrynchos.* This duck is the most abundant wild duck in the world and can be found from the polar regions through Asia, Europe and North America. Although common in California, it is less plentiful than the pintail duck.

The glossy-green head, narrow white collar and ruddy-brown breast of the drake or male make him impossible to mistake. The purple-blue wing patch or speculum is present on both the drake and hen although the hen is the mottled brown of so many female ducks. Both have a heavy-bodied conformation that makes them a favorite with duck hunters.

The wild mallard seems to adapt easily to many types of habitats. They are often found in the ponds of public parks and farms, and are the ancestors of almost all domestic ducks. In the words of Edwin Pickett, local ornithologist, ". . . it seems to enjoy the free-loading privileges and becomes quite tame". The mallard often crossbreeds with other ducks, and the offspring are fertile. There are some very interestingly marked ducks where they intermingle.

Many mallards stay in our valley and nest near the ponds and sloughs. The 8 to 10 eggs are incubated by the hen for about 26 days. When the young hatch they are taken to the water immediately.

Over the summer the adults lose the power of flight by losing flight feathers, and the male loses his distinctive plumage. By late fall, however, the power of flight and the breeding plumage of the male have returned.

If you would like to see the multitudes of waterfowl I have mentioned, visit Graylodge Wildlife Refuge or Sacramento Game Refuge during December. It is a visit you will not forget.

Pintail Ducks

The pintail duck, *Anas acuta* is possibly the most abundant waterfowl in California during the winter.

The Sacramento Valley yearly numbers may reach three million birds. It is also the most widely distributed duck in North America.

Pintails are slender-bodied ducks with slim, pointed wings and are, perhaps, the most easily identified flying duck in our area. Identification is made easy because of the long sharp pointed tail of the male.

The female's tail is also pointed but is not nearly as long. This characteristic gives the bird its scientific name—*Anas,* referring to the tail and *acuta,* meaning long sharp point.

During winter, the male has a rich mahogany-brown head and a snowy-white breast and belly. The female is a mottled brown all over and has a white rear edge to the usually brown speculum. The male somewhat resembles her at other times of the year.

Pintails or sprig, as sportsmen often call them, are surface-feeding ducks and are often called dabbling or puddle ducks. These are ducks that can leap out of the water with no running take-off. They feed on or near the surface of the water, principally on various types of water plants, sedges, grasses, rice, barley, and other cultivated grains.

Most of our pintails are migratory and may just pass through our valley. Others may spend the winter. Most, however, return north to nest. The first migrants arrive here about mid-September and many continue all the way into Mexico. In early spring, the migration north starts.

The Sacramento Wildlife Refuge operated by the U.S. Fish and Wildlife Service near Willows and Gray Lodge Refuge operated by our California Department of Fish and Game near Live Oak are world famous for their tremendous winter concentrations of waterfowl.

They welcome visitors, November and December are excellent months for observing.

Wood Duck

One of the real joys of nature may be missed by the casual observer through nature's window. A more active participation is usually necessary to get into the rather secluded habitat of the wood duck, *Aix sponsa*.

Although this duck is not uncommon, it does not visit yards or city lakes. Instead, they can be found feeding in woodlands where they eat whole acorns, berries, grapes, and a great variety of plant matter. They then retire to quiet stretches of streams, canals, or pools where they visit and preen themselves in seeming contentment. They may also roost in trees in the same area, and although apparently preoccupied, they are ever alert to any intrusion. When disturbed, they burst into the air with a distinctive cry of danger.

The male wood duck is one of our most beautiful birds and undoubtedly our most beautiful duck. The top of the head is a shiny green with a velvet crest, while a patch of white adorns the eye and chin. The back is brown while the wings are black, purple and white. The breast is brown while the belly is white. If that description doesn't make you want to see this bird, I am very surpised. The female has a white eye patch while her upper parts and breast are brown. Her throat and belly are also white

The nest of the wood duck is normally a cavity in a tree that is within a mile of water or right over the water. In checking the height of the cavity from the ground, I found a great dispartiy in them. One source said 6 to 30 feet, another 3 to 50 feet, and another 20 to 40 feet. Take your pick, but 3 to 50 takes them all in.

The female smooths out the bottom of the hole and lays her 10 to 15 smooth white eggs. She covers them with feathers plucked from her breast. After incubating them from 28 to 31 days, they should hatch. The hen flies to the ground or water the morning after all eggs have hatched. She checks the area for danger, and then calls the young. They must climb up from the bottom of the nest to the edge of the hole. They do this with the aid of very sharp toenails. Once there, they fling themselves fluttering to the ground. They are seldom injured. The female then leads them to the water. A mile must be a very long way for such short legs. What a way to start life.

Coot

If I were asked to mention one of the most humorous birds, I would have to rank the American coot, *Fulica americana,* in the top few. This very common water bird is familiar to almost everyone. It is found on almost all bodies of water from small lakes to the ocean. The very white bill, contrasting black body, white V under the tail and red eye make identification absolute. It is the size of a small duck, 13 to 16 inches, and is called by many names. The most common other name is mud hen.

The feet of the coot are not webbed as a duck's feet are. Instead, they have lobes on the sides of their toes. These serve very admirably for swimming. Actually, they are not related very closely to ducks at all but are in reality a marsh bird. They are classified in the Rail family.

In the winter time coots are often found in large flocks. They usually swim away from the shore when disturbed but usually do not fly. If they are surprised, they often run across the water with their feet and wings flapping. Their feet and wings are in rhythm and it looks like they can't take off. I have seen them do this for 2 or 3 hundred yards.

When swimming, the coot moves his head back and forth in a very humorous fashion. The faster he swims, the faster his head moves. They eat both plant and animal foods for which they dive or tip like the pond ducks. Before all of man's agriculture, they played an important part in keeping waterways open. They would eat the fresh new root shoots of the cattails. Their bill is very well adapted to cutting them off. Now, however, they like to graze on many of the crops man plants. When they congregate in large numbers, they are a serious pest to agriculture. They are often hunted for sport but are not generally desired for eating.

I can't help but admire this bird. He has met man and his progress head on and survived. I sometimes even have the feeling this wise old bird is laughing at us.

Sandhill Crane

I wonder how many people have heard the call of the sandhill crane, *Grus canadensis,* and not known what it was. We often hear their distinctive call coming from above the fog. Students will ask me what kind of a creature makes that noise. The call is variously described as: a low, loud musical rattle; a shrill, rolling garooo-a-a-a; a tuk tuk-tuk tuk-tuk tuk; and a deep, rolling crrronk. Take your pick. To me, they sound like a bunch of nervous raccoons.

The sandhill crane is one of two cranes found in North America. Much attention is paid to the much larger and nearly extinct whooping crane and its migratory movements. The sandhill is the only one we have in our valley, and it arrives here from its breeding grounds in eastern Oregon and northeastern California about the first of October.

While here, the cranes loaf, roost, and feed in the marsh and pasture lands. About 60 percent of them are found in the Sacramento-San Joaquin Delta. These areas provide roots, grain, bulbs, tender plants, frogs, mice and insects for the cranes' diet.

The sandhill crane is the larger of another race of cranes—the little brown crane. The little brown crane stands only about three feet high while the sandhill is four feet. Other than size, they are almost indistinguishable. The little browns breed in the Arctic and are very uncommon around here.

The adults of both races are entirely an ashy gray except for the featherless red-topped head. They also have short plumelike feathers that droop over the back. In flight, these fascinating birds hold their neck out straight in front and their black legs out behind. Groups generally fly in formation and characteristically flap their wings several strokes and then glide. Their wings show a very bowed appearance on the down stroke.

One of the more fascinating aspects of these birds is their courtship behavior. This starts before they leave in the spring and consists of bowing, hopping, skipping, and jumps high in the air. These antics may be performed by pairs or groups.

By mid-March the sandhills are all gone, but before this you still may see them flying overhead. Watch for them and see how you would describe their call.

California Condor

I enjoyed the privilege of participating in the 1973 annual California condor census when 36 individual birds were seen. Although the numbers of birds seen each year varies, the overall population of condors is generally considered about the same. The variations are probably due to weather conditions, food supply, fires, etc. Although it is all but impossible to take a complete census, those that study the statistics know that our condor is on the brink of extinction and, perhaps, the loss of only one bird could send the species into oblivion.

One often hears that a species may be doomed to extinction because of natural causes. Such is often said of the Sierra redwood as well as our condor. I hope that viewers through nature's window understand that time is a great variable. When one thinks of extinction caused by man it may come about in under a hundred years. Extinction by natural causes may be from thousands to millions of years. Projected natural extinction is under no circumstance an excuse for environmental destruction by man.

That these birds have been rare for a time is stated well by William Finley in a fine bird book called *Birds of America* published in 1917. He states, "If you were to start on a hunt for the California condor, you might search for years as we did without success." Ralph Hoffman in his book, *Birds of the Pacific States,* says, "The ambition of every California bird student is to see a condor." This book was first published in 1927.

The California condor, *Gymnogyps californianus,* has the largest wingspan of any North American land bird. They may weigh 20 pounds and have a wingspread of over nine feet. The adults are more than twice as large as the common turkey vulture, and the leading edge of the underwing has a broad band of white. Younger condors appear all black.

The Andean condor is a widely separated species found in South America. Although more numerous, it, too, must be considered an endangered species.

The last remaining California condors are found in the rugged mountains of Ventura and Santa Barbara Counties. Their continued existence will challenge man's ability to coexist on our planet with other forms of life. This is a real challenge man must not lose.

Marsh Hawk

I do not believe there is any group of birds that fascinate more people than do the hawks. Even those people that are not the least interested in most birds will not pass up an opportunity to watch a hawk on a fence-post or one soaring nearby in the air. A few of the hawks that are present in our area are very easily identified. One of these is the marsh hawk.

The marsh hawk, *Circus cyaneus,* belongs to a group of hawks called the harriers. This hawk has different colors for each sex. The female is a dark brown above with buffy white underparts. The male is a bluish-gray on the back and tail. The tail is banded with five or six dusky bands. The wing tips of both are black. The foolproof identification mark, however, is the snow-white rump spot present on both male and female. Once this is seen, it has to be a marsh hawk. Gulls, white-tailed kites, and some other hawks may have whitish tails but not rumps. Even young marsh hawks have a white rump spot so identification is definite for the whole species.

Both sexes also have the face encircled with an imperfect ruff somewhat similar to some of the owls. This gives them an interesting face which intrigues photographers.

The long somewhat rounded wings, long narrow tail and wings held slightly angled are some of the field marks used by bird watchers to identify the marsh hawk from a distance. The flapping sailing flight, swinging slightly from side to side at almost grass level is characteristic of the hunting flight. They are generally found in open country and feed on squirrels, rabbits, lizards, snakes, frogs and evidently some birds. Their food is undoubtedly dependent on the season and locality. They nest on the ground and start incubating when the first egg is laid. There are usually four to six eggs

and the younger birds are sometimes not able to compete with those first hatched.

We usually have an influx of marsh hawks in fall and winter. They come to us from the north and the mountains and supplement our resident birds.

This hawk, like all hawks and owls, are now protected by both state and federal laws. It is illegal to injure, kill, or possess any of them, even the fledgling birds, nests, or eggs. If you ever see anyone shooting or molesting hawks or owls, call your local California State Fish and Game warden. He knows what to do.

Red-tailed Hawk

Of all the scenes through nature's window there is one that always evokes the same feeling in all of us. The lazy soaring of a hawk gently spiraling upwards on the rising air currents creates, in all viewers, the yearning to be as free as a bird.

The hawk that we most often see soaring in our valley is the red-tailed hawk, *Buteo jamaicensis.* It is a beautiful, majestic bird of prey that is often seen perched high in trees, on high voltage lines, or on telephone poles. As is common with birds of prey, the female is larger than the male although both are quite large. The body of the male is about 22 inches, while the female's is 24 inches. The wingspread is 4 to 4 1/2 feet with rounded tips.

Adults of this species have zoned markings on the underparts with a brownish ground color above. The rounded tail often appears whitish from underneath although strong sunlight shining through may give a hint of the bright rust color on the top of the tail.

Young red-tailed hawks have a dark gray tail that may or may not be banded. The belly is basically a light color with large brown splotches. The breast usually has a clear white area that makes a good field mark for perched immatures.

There are many variations in the coloration of hawks. They may be very dark so that most field marks are obliterated or they may be very light so there are no field marks. This makes hawk identification very difficult. Even the most skillful bird watcher cannot identify every hawk seen. However, when a red tail is seen on a large soaring hawk, you can be sure it is a red-tailed hawk.

Animals such as ground squirrels, mice, rabbits and reptiles including rattlesnakes serve as food for the red-tail. It is possible that barnyard fowl may be taken, but few incidents have been verified. Nevertheless, the term chicken or hen hawk is sometimes mistakenly applied to this bird. This sometimes serves as an excuse for unsportsmanlike hunters to kill them. It should be remembered that all hawks and owls are protected by law in California, and killing, injuring, or possessing them is illegal. This valuable environmental partner should be enjoyed and protected by all of us. If you see someone shooting hawks, call the California State Department of Fish and Game. Do your part to protect our wildlife.

Sparrow Hawk

I feel quite strongly that it is not appropriate to attach "good and bad guy" roles to nature's creatures. This concept, however, does not preclude a person from having favorites. Just as some people prefer dogs over cats, it is also possible to have favorites in the wild kingdom. One of my favorites is the beautiful and graceful little sparrow hawk, *Falco sparverius.* It is also more properly called the American kestrel.

The sparrow hawk is the smallest and by far the most common representative of the falcons we have in our valley. Falcons have long pointed wings, large heads, and tails

that narrow at the tip. They have a fast direct flight even though the wingbeat may be somewhat choppy.

The sparrow hawk is the American representative of the European kestrel and is a bird of open spaces like fields, meadows and plains. He must have perches, however, and in our valley he especially likes powerlines and poles along the roads. He can be seen there frequently especially through the winter when our resident population is increased by those moving in from the north and mountains.

The adult male has bluish-gray on his wings, a bright rusty back and tail, and black-splotched buffy underparts. His tail has a broad black band near the tip. The larger female is more brown, has a heavily barred back, and five or six black bars in her tail. Both male and female have a grayish cap with rufous in it, and two black stripes down each side of the head—all of which makes this little hawk unmistakable. There is another thing to watch for, however, and that occurs when the sparrow hawk lands on his perch. Whenever he does, he will wag his tail a few times.

This slightly larger than robin-sized hawk is a real opportunist. He eats anything that is plentiful like lizards, grasshoppers, mice and small birds. While hunting, he hovers frequently and is quite distinctive. The white-tailed kite is the only other common hawk that hovers so frequently.

Barn Owl

A flash of white in front of the car lights at night is often all people see of the barn owl, *Tyto alba.*

The whiteness leads many people to believe they have seen a snowy owl, but the snowy owl is very uncommon in our valley.

The barn owl is a medium-sized owl with a wing spread of up to 45 inches.

The white face is heart-shaped and rimmed with brown. The eyes are small and quite dark.

There are no ear tufts. The back is a light golden-brown.

The undersides are white with varying amounts of dark flecks.

Barn owls like to sit on fences and telephone poles along country roads. As mice and other rodents cross the roads, the owls swoop down and grab them with their claws.

We often see them in our car headlights. Many barn owls are hit and killed.

Our area has one of the largest barn owl populations in the United States. Quite often, Sacramento has the highest barn owl count on the annual Audubon Christmas Bird Census.

Originally, the barn owl nested in cavities of cliffs, gullies, or large trees.

Now it uses attics, abandoned buildings, steeples and and, of course, barns.

They generally hunt over grasslands and alfalfa fields but also use roads, as mentioned previously.

Their flight is silent because of their soft plumage and special wing feather construction.

Food consumption of the barn owl is enormous and food equal to their own weight is consumed every day if possible.

Usually five to seven white eggs are laid in the nest. Each egg is incubated as soon as it is laid.

Because of this, there are different sizes and ages of young in the nest.

All hawks, owls, and almost all other birds are now protected by law. They cannot be killed, injured, or possessed, dead or alive, without a special permit.

Burrowing Owl

Owls seem to hold a special fascination for many people, even nonbird watchers. I have been in many homes where replicas of owls serve as ashtrays, incense burners, vases, candle holders, pictures, etc. The use of owls in decorations is quite common and popular. The live owl, however, is far more intriguing. One of the most interesting is the burrowing owl, *Speotyto cunicularia.*

The burrowing owl is a small light-brown and white owl that is 9 to 11 inches long. The small head is round with yellow eyes and no ear tufts. The most obvious difference between the burrowing owl and other owls is the rather long legs that are unfeathered below the knees. The fact that he also hunts in the daytime is rather unique. However, most hunting is done at dusk.

The primary food of the burrowing owl is rodents, lizards and snakes, but large insects like crickets, grasshoppers, and dragonflies, are taken on the wing.

Naturally the burrowing owl nests in burrows, and in our area, he probably does not dig his own. The burrows of ground squirrels, foxes, badgers, skunks, etc., are used although the owl may enlarge or alter them.

The large brood of 8 or 9 young require an enormous quantity of food and evidently keep the parents hunting day and night. Often one of the parents can be seen sitting quite near the entrance to the burrow. It is usually possible to approach him quite close as the owl seems to depend upon his protective coloration to conceal him. As you move, the eyes and head of the owl follow you. Folk tales say that if you keep walking around, you can make him twist his head off. Naturally he doesn't do it, but the owl can turn his head around so fast you miss it turning.

This fascinating little fellow is also called the ground owl or billy owl. Watch for him in open fields and on fence posts. I am sure you will feel like calling him Billy too.

Crow

Few birds in America need as little introduction as the common crow, *Corvus brachyrhynohos.* This large, chunky, black bird when full grown is 17 to 21 inches from the tip of his bill to the tip of his tail. The crow is completely black including his bill, legs and feet. A violet metallic sheen may be seen in strong light on the back while the wings may appear bluish green.

The only bird that one could easily confuse with the crow is the common raven, *Corvus corax,* which is larger and may reach 27 inches. In flight, they both flap and soar. In soaring, the wings of the raven are held flat, while the wings of the crow are slightly elevated. Another excellent characteristic for identification is the tail shape. The raven has a wedge-shaped tail, while the crow's tail is slightly rounded. Size alone should eliminate confusing the crow with the Brewer's blackbird. This much smaller bird is only 10 inches long.

It would be quite unusual to find a raven in our valley although the crow is quite common. You may see ravens along the coast, however.

Our crows are not migratory in the true sense of the word, but they may descend out of the foothills in winter or even move to other more suitable areas. They are very partial to agricultural areas and as is well-known, may be a real pest. Except during nesting, they often accumulate in large roosts for the night. There are a few of these in Sacramento proper and several near by. Most of the year they feed on injurious beetles, cutworms, wild berries and the like. Their brief periods of crop depredations have earned them an unsavory reputation, and they have been persecuted by man for many years. Man's attempts to control the numbers of crows has not been effective except in some localized instances.

The crow's organized flock behavior and high intelligence has protected the species very well.

Yellow-billed Magpie

Many of the characters one sees through nature's window are so active and obvious that little description seems necessary. They lead colorful and exciting lives and play important roles in nature's scheme. In our valley the yellow-billed magpie, *Pica nuttali,* is one of the most fascinating of the obvious characters.

This handsome bird is about 18 inches long although his long, black, wedge-shaped tail makes up about 10 inches of his length. His markings are as bold as his actions with his pure white belly and broad white wing splash. There is not another bird in our valley that can be confused with the yellow-billed magpie. His black and white pattern is

identical, however, to the widely distributed black-billed magpie found in the Great Basin. Their main difference is as their name implies—the color of their bills. The yellow-billed magpie is found nowhere else in the world but in the central valley of California.

Their bulky stick nests are a familiar sight to valley residents. They are seen in the tops of our oaks, and several of them may be in one tree. They are usually roofed and have two entrances.

Magpies' food consists mainly of insects, particularly grasshoppers, beetles and hymenopterans (wasps, ants, etc.). Large amounts of carrion and spiders are also eaten. Some plant foods are eaten such as oats, wheat, grapes and nuts. It is their crop stealing that gets them in trouble with farmers. However, most who have studied this bird are convinced his beneficial insectivorous habits far outweigh his crop depredations. It is partly because of this that the yellow-billed magpie is protected by law.

Magpies are members of the crow family and as such are very wary and intelligent. They have been able to adapt to man's intrusions, and as long as there are open fields to feed in and trees to nest in they seem to survive. Watch this noisy, nosey neighbor of ours, and I am sure you will enjoy his fun-loving antics. He is definitely one of our most interesting characters.

Goldfinches

It is always surprising when we see something that we do not expect. The lawns of our homes, schools and parks are frequented by bouncy flocks of robins, starlings, magpies and several other visitors looking for free chow. We get rather used to seeing these larger birds in these places but often miss the tight little flocks of goldfinches earnestly searching out seeds from between the blades of grass. These sassy little fellows are fascinating residents of our area that travel in closely knit flocks that do not seem to spread even when they fly. They are with us throughout the entire year so we get a chance to see them in their best "bib and tucker."

There are actually two kinds of goldfinches common here and learning to tell them apart can be fun. They both have yellowish underparts and black wings with white wing patches. The male American goldfinch, *Spinus tristus,* in summer has a perky little black cap that is lost in winter along with his yellow back. Both the male and female have a whitish rump spot and look very much alike in winter. The male loses his bright yellow back and turns an olive color very similar to the female.

The other goldfinch is the lesser goldfinch, *Spinus psaltria,* and although they are slightly smaller, it is difficult to tell them apart by size. The male keeps his black cap during the winter and summer while the female does not have a cap at all. Their backs are a dark olive color and they have no whitish rump spot.

As seed eaters the goldfinches are unsurpassed, and they are especially fond of thistle seeds. When star thistle, chickory, dandelion, or milk thistle start producing seeds, flocks of these birds are never far away. In spring when you see them in their beautiful breeding colors, you will see why some people call them wild canaries.

Hummingbirds

"Glittering fragments of the rainbow" were the words used by one of our greatest naturalists, John James Audubon, to describe the hummingbird. The name hummingbird was applied by the first North American English colonists when they heard the hum of its wings. There are 319 kinds of hummingbirds in the world, and most are found in the tropics. There are 15 found throughout the west but only five regulars in northern California. The smallest bird in the world is the bee hummer of Cuba. It is only two and one-quarter inch long and weighs no more than a thin dime.

The most common and only resident hummer found in our valley is the Anna's hummingbird, *Calypte anna.* The male Anna's is the only United States hummer with an iridescent red crown. The throat is also an iridescent red and called a gorget. The red shows only when the light is reflected at the right angle. It is then dazzling and justifies a name often used for them—winged jewels. The back and rump and tail are green while the dusky underparts are greenish tinged.

Identifying female hummers in our area is very risky. Even experts have great difficulty distinguishing the various bronzy green females. Females select the nest site and proceed to build the nest by themselves. The female uses the down or fuzz from various plants and holds it together with spider webs. Her two eggs are bean size which she alone incubates. Generally, the male leaves after mating and plays no other part in family life.

Hummers are the most maneuverable of all birds and can fly forward, backward, up, down, sideways or hover.

The wings of most hummers beat at the fantastic rate of 40 to 80 times a second and may beat 200 times a second during a power dive. They feed every 10 to 15 minutes in order to sustain their high metabolism. It has been estimated that the food requirements of a hummingbird is equal to a 170 pound man consuming 140 pounds of bread or 280 pounds of boiled potatoes a day. This breaks down to 155,000 calories a day. (Almost as much as a teenage boy.)

They feed on flower nectar and various insects. They can be brought in quite close with hummingbird feeders that can be made or purchased in pet stores. A mixture of one cup of sugar to nine or ten cups of water and a few drops of red food coloring will do fine. Honey should never be used. If the sugar does not dissolve readily, boil the mixture. Your efforts will be rewarded, and you may become confirmed hummingbird watchers as some members of my family are.

Hummingbird

House Finch

The enjoyment one can experience when a flock of bouncing, crimson bits are seen in the yard is one of the pleasures reserved for those who love nature.

The tiny jewels I am referring to have a variety of names. California linnet or house finch are most often used, but they are also referred to as red heads.

Their scientific name is *Carpodacus mexicanus.*

The birds are the size of a small sparrow with a total length between five and six inches. They are very closely related to the sparrows. The tail of both male and female is square ended.

The adult male is the colorful one with a bright crimson forehead, rump, chin, breast and a line over the eye. The belly is whitish with distinctive brown streaks that distinguish it from the male purple finch that visits our valley in the winter.

The back, wings, and tail are brown.

The female has no crimson color at all, but is grayish-brown above and whitish, streaked with brown, below. She does not have the white eyeline that the very similar female purple finch has.

In spring, the color of the male house finch is quite variable. I have seen some males that are of the color of a new penny to pale yellow. Do not let these variations fool you, they are still house finches.

No other western bird has adapted in such large numbers to the changes man has made in the environment.

The female builds the nest alone and is quite adaptable in selecting a site. The nest is seldom more than seven feet from the ground and may be in a vine, rosebush, light fixture or shrubbery.

Weed stems, grasses, string, rags and many other items may be used to build the nest. The female incubates the four or five eggs while the male feeds her. Two broods may be raised in the same nest during the season.

Although the main diet of the house finch is weed seeds like filaree, turkey mullein, mustard, star thistle, etc., they also like cultivated fruit.

In large numbers they may be a serious pest to farmers.

Their cheerful, melodic call and bright color, however, make them enjoyable to have around. Watch for them.

House Sparrow

In 1850 the house sparrow, *Passer domesticus,* was introduced into America. Although the little fellow has been called English sparrow for many years, the bird watcher has changed its name. Eight pairs were imported in the fall of that year and released in the spring. There were undoubtedly other introductions, but the reproductive ability of this hardy little immigrant probably made them unnecessary.

The house sparrow can raise three or more broods a year, and each brood consists of three or more young. The parents are surprisingly successful raising their brood and can populate an area with amazing rapidity.

Of all birds one may see, the male house sparrow is probably the best known. The gray cap, black throat and black bill of the male cannot be confused with any other bird. The female, however, with her grayish-brown head, streaked buffy back and whitish underparts can be confused with many others. The house sparrow is about six inches long. It is not a true sparrow but a weaver finch.

The original home of our pesky little friend is Europe and the British Isles. It is now widespread in the United States and found anywhere man lives. He inhabits cities, suburban areas and farms but is seldom found in the wild.

The house sparrow was originally imported because of his seed-eating characteristics, especially the seeds in horse manure. Horses were soon discarded; however, the bird remains. His aggressiveness and gregarious behavior have not made him a favorite addition to our bird world. The droppings of large flocks discolor buildings, their nests often clog drains, and native birds are intimidated and often driven out.

Considered in its own right, however, one must admire this fiesty feathered foreigner. I once saw a cartoon of the trials and tribulations of a pair of "English sparrows" called Monte and Gwendolynne. Since that time, I have looked upon this bird in a different light.

Gulls

One of the most obvious birds in our area is probably the gull. Notice I said gull, not sea gull because the name sea gull belongs to a specific kind of gull, generally called the herring. It is not very common in our valley. Perhaps I should not put so much emphasis on the difference, but to be correct the name of the group is gull. Using the name sea gull for all of them is like calling all dogs police dogs. It just isn't right.

There are two common gulls that make up almost all of the gull population that we see. One is the California gull, *Larus californicus,* and the other is the ring-billed gull, *Larus delawarensis.* Both are very similar in appearance and certain field marks must be seen to differentiate them. The adults of both have white heads, tails and underparts. The back and top of the wings is a light gray although the gray is darker on the California gull. The difference most of us use, however, is the marking on the bill. The ring-billed gull has a black ring near the tip of the bill while the California gull has a red spot on the lower bill. Usually, a complete or incomplete black band also appears in front of the red spot. Leg color can also be used since the California gull has greenish legs while the ring-billed gull has yellowish-green legs. The wingtips of both are black with white spots.

It is more difficult to tell immature gulls apart and I won't say much except that young gulls are very dark brownish and get lighter each year. The third year most attain the adult plumage. Second year ring-bills have a narrow black band at the tip of their tail so they can be easily told.

Food of gulls found inland is quite varied.

Garbage dumps draw them in large numbers as does freshly plowed ground. A lunch sack in an empty parking lot is an irresistible lure for almost all.

Gulls have long been protected because their scavenger abilities have been recognized for a long time. Their gentle and graceful flight is admired by all and they are an interesting bird to observe. Can you tell these two common gulls apart?

Scrub Jay

When one looks through nature's window there are many scenes that meet the eye. They vary from backyards to wilderness visits. In all of the settings of our valley and foothills the scrub jay, *Aphelocoma coerulescens,* will be seen.

Notice I said scrub jay and not blue jay. The blue jay, *Cyanocitta cristata,* is an eastern species of jay that, although blue, is quite different from our bird. When one applies the term blue jay to our local species, it is like calling a beagle a shepard.

All jays are members of the crow family, and as such have dubious reputations. Nevertheless, one cannot help but appreciate their boldness and their straight-forward approach to life. There is no sneaking or skulking around for these characters. When they raid your fruit trees they do so in rowdy gangs and make plenty of noise in the process. They will live or visit in your yard all year and loudly protest any activity in an area they consider theirs; that is, except in spring when they are nesting. During this short period they become strangely quiet near their nests.

The nest is a bulky structure of twigs, grass, and moss that may be placed in any available tree. The nests are difficult to see in spring and summer when leaves are present.

The scrub jay is one of the few jays without a crest. The head, wings and tail are blue while the back is pale brownish. The underparts are pale gray with a whitish

throat. The overall length is about 12 inches, and there is really no other bird in the area it can be confused with. They eat anything that is in season such as fruit, insects, eggs, and nuts. They can often be seen holding an acorn or almond down with a foot and pounding it woodpecker style with their beak. This versatile, adaptive and intelligent bird makes an interesting subject for a study. Watch and see.

Killdeer

The nesting season for many birds is an especially dangerous time. They must stay in a rather confined area and the search for extra food for the young is greatly intensified. Every spring I have several baby birds brought to me that people have found on the ground. Most people assume they have fallen from a nest, but one of our most common and obvious birds builds its nest on the ground. It is often these young that people find. I am referring to a bird familiar to almost everyone—the killdeer, *Charadrius vociferus.*

The species name vociferus applies quite well to this pleasantly noisy bird that cries "killdeeah" so insistently. Its two bold black breast bands, golden red rump, and striking white wing stripe make field identification quite easy and positive. The distinctive head pattern and rather long legs and tail just add to the unique patterns that set this bird apart.

The nest of the bird is little more than a small scraped area in a place with good visibility. They will use bare ground, a bald spot in a pasture, gravel on a shore, roadway, or even a roof. The four eggs are buff colored with dark spots. Both adults will protect the nest by feigning a broken wing and crying piteously. They use their act to decoy a possible enemy away from the nest. When the intruder is far enough from the nest, the decoying bird suddenly recovers health and bursts into flight and song.

The young are ready to follow their mother within two hours after hatching. They may get separated from their mother by dogs, cats, or people. Mother and father, however, can fly high to look for their offspring.

If you find the young they should be left in a safe place on the ground for their parents to find. Raising them is quite difficult and usually leads to their death despite our good intentions. This is a sad commentary for a good samaritan.

Western Kingbird

For those who are sensitive to the creatures of nature, enjoyment can be found everywhere. Places that would seem to be barren of animal life often provide observers through nature's window a very interesting scene.

In summer, when you travel the back roads through the dry grassy rolling hills surrounding Sacramento, you will probably see an interesting bird that comes to our valley to raise its family. One of the best places to watch is the top two strands of barbed wire that parallel the highway. I am referring to the western kingbird, *Tyrannus verticalis.*

These handsome flycatchers usually arrive in April, and pairs will soon start their territorial homesteading. During the height of

the mating season, the male does a series of jerky aerial maneuvers accompanied by a continual twittering that usually starts before dawn and continues all day.

The western kingbird has a head, neck, back, and breast of ashy gray. The wings are brown, and the belly is a striking yellow. The tail is coal black with white edges, and the male has a concealed crown patch of scarlet. In all, an easily identified bird.

Typical behavior of this common species is to sit on a perch and watch for insects. When an insect is sighted, the flycatcher darts out, catches it and returns to the original perch, or perhaps one nearby. The kingbird will often capture grasshoppers on the ground. He is also noted for his courageous behavior against larger birds and will not hesitate to attack birds such as crows, owls, or herons that enter his territory.

Do not look for this bird in the winter, however, because in August they begin leaving us. The rest depart in September for their winter quarters located near northern Mexico all the way south to Nicaragua.

Watch for the western kingbird when you travel. When you find one in the Sacramento Valley you can be sure summer is here.

Mockingbird

Even when winter is still in its strength, nature presents many spring previews to those who care to watch.

If one looks closely at the branches of trees and bushes, the sleeping buds show signs of stirring from their winter's rest. Tiny bits of leaves can be seen peeping from beneath the bud covers indicating the plant world is ready and waiting to burst forth in spring. Perhaps these subtle signs cause an

awakening of desires in members of the animal world. Observing these indications can bring us great pleasures.

Even when the nesting season has not arrived, several species of birds begin to exhibit a territorial urge. The one most obvious to me is the western mockingbird, *Mimus polyglottis.*

This handsome robin-size bird has a long black tail brightened by white outer-tail feathers. The white wing patches are quite obvious against the overall grayish color. The underparts are a lighter gray and there is a light stripe starting at the bill and running through the eye. The bird appears more slim than a robin and his long tail is quite expressive.

I am very fond of the mocker that frequents my yard. In summer he eats my grapes and catches insects. In fall and winter he eats berries and anything else he can find. His territorial behavior in late winter is quite fascinating. He has no mate, no nest, and as far as I can tell, no competition. Still he flies at all the other birds with his wings spread in a threat display and sits in the leafless trees looking as fierce as he can. Though he has not burst forth in his fabulous song imitations, I know they will follow soon.

I am sure all of you have at least one mockingbird in your area. All this striking fellow needs is a few trees, shrubs and lawns for his larder. I am sure your mocker won't be as impressive as mine but try watching yours and see.

Loggerhead Shrike

There are many incongruities to puzzle the observer through nature's window. None, however, is more perplexing than the songbird turned hunter. I am referring to the loggerhead shrike, *Lanius ludovicianus.*

This unique bird is related to thrushes and other song birds but is adapted for a predatory life. The black mask through the eye, its black wings with white patches, black tail sporting flashy white outer tail feathers, and dark grayish back with light underparts distinguish this resident from all of our other birds. The only somewhat similarly colored bird is our mockingbird.

The loggerhead shrike gets the name loggerhead from its bulky head. The shrike part evidentally refers to its call. The silhouette of this bird is very distinctive, and once you become familiar with it you can recognize it easily. It is often seen on the wires along

roads, fence posts, stakes in open fields or on the ends of tree branches. These vantage points are used for hunting.

The food of the loggerhead is almost all animal. Lizards, grasshoppers, beetles, and other birds are typical types of food. The feet of the shrike, however, seem to be inadequate to tear up larger prey. Because of this, they often impale their catches on thorns, ends of sticks or the barbs of barbed wire. This enables the shrike to eat its prey in small pieces by pulling it apart. This habit has earned the shrike the nickname "butcher bird." The weakness of the feet is further indicated by the way shrikes usually carry their catch in their beaks. The beak is also used to dispatch anything it catches. Actually small birds comprise a very small part of their diet, and informed bird watchers recognize the values of this ecological partner.

Increased knowledge of ecological factors has caused the change in status of this interesting bird from unprotected to protected. It is comforting to know we have made some progress in understanding our environment.

Rock Dove

It seems peculiar that one of the most obvious partners in our ecological scheme is almost ignored by those skilled in nature. I am referring to the rock dove or domestic pigeon, *Columba livia.*

This common city dweller provides millions of people all over the world countless hours of enjoyment. Their peaceful ways, pleasant gurgling sounds and affinity for man endear them to the many people that frequent city parks. They make pleasant companions for those that do not have signs, buildings, and statues affected by their nests and droppings.

Since the rock dove is not native, it is often ignored by bird watchers, and it will surprise you to learn that many bird and nature books do not even mention it exists. The rocky coast of Europe was evidentally its original home, but it is now found throughout the world. Man was probably the principle cause of the pigeons' spread. He used the pigeons and their homing ability to carry messages. Over the years many diversified strains have been developed. Characteristics of their strains often show up in the feral or wild flocks we see. In fact, the white, tan, gray and blackish birds in the flocks make identification easy. Our only other pigeon, the band-tailed, has uniformly colored flocks and each bird has a pale broad band across the end of his fanlike tail. It is never found in cities.

The typical rock dove has a white rump, broad black band on tail, two black bars on the wings and red feet. Their flight speed is about 67 miles per hour, and they may live 16 years.

They are now considered a game bird, and according to fish and game regulations, may be taken only during the band-tailed pigeon season.

The rock dove is a real enigma as it is ignored by bird buffs and revered by others. People are really fickle.

Band-tailed Pigeon

When most people think of bird migration, they think of birds flying north or south in flocks on a rather regular schedule. Such is not the case of the band-tailed pigeon, *Columba fasciata.*

Although the bandtail is a migratory bird, its habits are so erratic that it is difficult to predict where or when it will appear next. It prefers to live in natural forests. In summer they move to higher elevations. Their

breeding range is mostly the north coast range and higher elevations of the western slopes of the Sierra Cascade Mountains. They are present in our valley October through April and sometimes appear in large flocks concentrated in certain areas for no easily explained reason. If you travel in the fringe or wooded areas of the Sacramento area, watch for flocks of bandtails.

Although the band on the tail that gives them their name is not obvious without the tail being spread, they can be easily identified. It is interesting that some authors mention the broad light gray band on the tip of the tail, while others mention the slender black band above the gray. Most also have a white crescent on the nape of the neck with metallic-colored feathers below. About the same size as a rock dove or domestic pigeon, a bandtail has a dark-tipped yellow bill and yellow feet. The rock dove has a white rump, black band on the tail, darker bill and red feet. The mourning dove has a pointed tail with large white spots.

Bandtails are fruit and seed eaters, and acorns make up the bulk of their diet. They prefer blue and California live oak acorns. Although none of my references mentioned it, they also eat ripe olives in great quantities. I have seen them with so many in their crop they could barely fly. I have often wondered how they tasted after a diet of olives and acorns. Maybe this is why they did not follow the same fate as the passenger pigeon. I have heard, however, they are a fine eating bird when properly prepared. They were saved from extinction however, by game laws, not by poor tasting flesh.

Watch for flocks of pigeons that are a uniform color. They are undoubtedly bandtails.

Mourning Dove

Late summer for observers through nature's window can mean many things. The grasses have turned brown, the mornings are crisp, and the seeds of turkey-mullein are ripe and attract the mourning dove.

The mourning dove, *Zenaidura macroura,* is a year-round resident of the Great Valley and one of the favorites of all who study our environment. His brownish body, bluish-gray head top, and very long central tail feathers that taper to the edge are a familiar sight to both motorist and bird watcher. This gentle creature is one of the most common of the larger birds that perch on the power lines along roadsides. His small head, broad shoulders, and body that tapers to the long tail easily distinguish him from the sparrow hawk and loggerhead shrike that also perch on the power lines.

There are only two other members of the dove group in our valley: the native band-tailed pigeon and the introduced rock dove or domestic pigeon of Europe. The long white-tipped tail feathers of the mourning dove quickly identify him from the pigeons with rounded tails. The mourning dove is truly an American bird since it nests in every state. It is so adaptable that it is equally at home in the western mountains, the open prairies, and the eastern woodlands. There are two subspecies recognized—an eastern and western.

The food of the mourning dove varies from one locale to another but is essentially

Mourning Dove

plant food. In our valley turkey-mullein is the single most important plant although seeds of California poppy, star thistle, mustard and wheat are also important. Mourning doves should not be encouraged to come to bird feeding stations in any numbers. A disease can be transmitted that can be fatal to them as well as other species.

Watch for this dove and listen for the whistle of his wings in flight. He presents an image of gentleness and has long been a symbol of peace—a commodity much in need today.

California Quail

A startling, sharp, sudden whir of wings is often one's introduction to the California quail, *Lophortyx californicus.* Their powerful flight muscles move the short wings so fast that they make a noise similar to an airplane propeller. This is a common characteristic of the heavy-bodied, chickenlike gallinaceous birds which include pheasants, grouse, turkeys, partridges and, of course, quail. They all have a short heavy bill, short rounded wings and rather long legs. They are capable runners who forage on the ground for seeds and insects.

The California or valley quail, as it is sometimes known, is the official bird of California. It is not migratory and seldom moves far from the warm lowland or brushy foothills where it makes its home. It is also

found in Oregon, Washington, Nevada and Baja.

The male has a black face that is bordered with white. His chest is blue-gray, his flanks are gray with white marks, and the top of his head is dark brown. The buffy feathers of his upper belly are edged with black which gives them a scalelike appearance. There is a rich brown patch in the middle of the belly. The forward drooping, comma-shaped black plume is, perhaps, his most prominent characteristic.

The female is chiefly brown with her belly and flanks similar to the male. Her plume, though similar, is not as large as the male.

The female lays her eggs in a depression in the ground that is lined with grasses and is located under a bush, log or rock. The eggs are a creamy-white and spotted with light golden brown. She may lay anywhere from six to 28, but 13 to 17 is the usual number. They have so many natural enemies that broods of a dozen or more are necessary for survival. The young leave the nest immediately and are soon on their own scratching for food. They remain in coveys of two or more families until the following spring when they disperse.

Undoubtedly, the California quail is one of my favorite birds. How about you?

Cliff Swallow

No story of spring would be complete without the mention of the return of the swallows. This romanticized event is familar to almost everyone. The swallow most often referred to is the cliff swallow, *Petrochelidon pyrrhonota.*

This sparrow-sized bird has the characteristically pointed wings of the swallow but is the only swallow with a square tail. The creamy-white forehead, blackish back and light-brown rump spot make identification positive. The gourd-shaped mud nests are built in colonies under eaves, bridges, etc. Where these places are not available to them, they nest on rough cliffs.

Food of this fascinating summer visitor is evidently 100 percent insects. The quantity consumed by a breeding colony is tremendous. Nestlings are fed close to their own weight every day while adults may eat nearly half their weight. Insects are generally caught while on the wing. The beneficial aspect of this species should not be underestimated, and they should be protected at all cost. They are, of course, protected by state and federal law. The nests also come under legal protection as soon as an egg is laid in them.

The cliff swallows start arriving in central California in February although the peak of the migration is in early May. They spend their spring and summer in nest building, incubation, and preparing the young birds for the long migration to the remote parts of Brazil. They all leave the area by the end of September.

At San Juan Capistrano, legend has it that they always arrive on March 19th, St. Joseph's Day. Birds that arrive early are either ignored or called "scouts". Natives, however, are hard pressed to define the purpose of the scouts since they certainly do not fly back to South America to warn the other birds. According to legend, they leave Capistrano on October 23rd, San Juan's Day.

This migration was made famous by the song "When the Swallows Come Back to Capistrano." The swallows there are no different than the ones I used to band under the culverts of Highway 50. No one, however, has immortalized them by a song titled "When the Swallows Come Back to the Culverts of Highway 50." Too bad!

Cliff Swallow

Purple Martin

In my opinion, there are very few attributes that other parts of the United States have that California does not. One thing we do miss in California, however, is the ability to attract purple martins, *Progne subis,* to our yards by building fancy martin houses.

I remember as a young boy in Ohio, the beautiful and fancy birdhouses suspended on a pole in the gardens. These houses would often attract colonies of almost a hundred birds. This influence is often expressed in the West by the mock birdhouses that decorate our own houses today. The purple martin that we have here is the same species as that found in the East. In the West it does not nest in houses for some reason. Instead, scattered pairs nest under eaves, cornices and ledges of various types in cities. A few pairs nest in downtown Sacramento which is one of the few nesting colonies known in central California.

The purple martin is the largest of the North American swallows. The male is purplish-black all over, while the female is purple only on the back. Her breast is brown with the rest of the underparts gray to white. They both have long wings and a deeply-forked tail. They usually glide in circles with alternate flapping of their wings. They spread their tails often.

The first purple martins arrive in Sacramento sometime during March, and they leave during August. They are usually obvious during this period along 9th Street from J to L streets where they perch on antennas and dart out to catch food. They consume many, many insects as do other swallows.

Their winters are spent in the Amazon Valley of Brazil. Watch for them on summer days and see if you can discover what kind of housing Sacramento's purple martins prefer.

Purple Martin

Roadrunner

The roadrunner, *Geococcyx californianus*, is a fascinating bird seen in cartoons, commercials, decorations and jewelry so often that the actual creature seems unreal.

Although the roadrunner reminds me of a hen pheasant, it is really quite different. It is about 24 inches long with a mottled brownish color, shaggy crest and orange bar behind his eye. The long expressive tail is raised and lowered with jerky movements. The tail also shows white on the edges. The feet may have 2 toes pointed forward and 2 backward. The legs and feet are a pale blue.

The roadrunner is generally associated with the southwest deserts, but its range extends clear to northern California. They have been seen around Folsom, Shingle Springs, and Rio Linda and they undoubtedly exist in may other areas.

The snake and lizard-eating characteristics of the roadrunner are well-known. Lizards are run down and caught with deft movements. Even rattlesnakes are caught and eaten although the roadrunner is not immune to the venom.

In times past, the roadrunner was often seen running ahead of a horse and buggy. Today, however, the 18 mph of the roadrunner scarcely equips him to race our modern cars. Perhaps this is why we see so very few.

The roadrunner is a member of the cuckoo family and fossil forms have been found in the LaBrea Tar Pits. Its peculiar habits have endeared it to many. It is the state bird of New Mexico and a favorite symbol in Indian jewelry.

Roadrunner

Robin

It is always interesting to hear about the things that people see through nature's window. Those of us that are constantly aware of the happenings in nature often do not realize others may wonder about things that we take for granted.

Many times I have been asked about the large number of robins in the Sacramento area in winter. The number of robins always increases in fall and winter and their numbers are not too difficult to explain.

The difference between fall and winter in the valley is difficult to discern, although in winter the major rain storms usually occur and all the leaves have fallen from the deciduous trees. Tule fogs are common and extended periods of cold weather occur. The days are often sparkling clear and the mountain snow packs can often be seen in the distant Sierra. Few, if any, plants bloom in our gardens and most lawns have extensive brown areas due to the weed grasses that infest them. Nature seems to be taking a rest to prepare for the tremendous surge of spring. All that is left for the creatures of the animal world is the remains of the summer's productivity.

Robins, for instance, are forced down from the Sierra slopes when the mountain berries are covered by snow. The farther down the snow pack reaches the more birds are concentrated in the valley. Once here, they tend to disperse and take advantage of every available berry. They usually stay till spring.

During the colder nights birds such as robins often accumulate in roosts that consist of a few isolated trees to large groves. The birds prefer the evergreens like the live oaks for they find warmth when they settle into the inner depths of their branches. Their accumulated body heat insulated by the leaves help keep out the chill of winter.

Another characteristic of winter that one may miss entirely is the small green plants that unobtrusively appear between the dead remains of last year's growth. As winter progresses, these tiny, almost invisible, plants grow and slowly cover the valley with a thin blanket of green. The first few seeds that germinate are indications of nature's fantastic promise of spring.

Varied Thrush

One of the many pleasures that the viewers through nature's window enjoy is the constant variety presented. Not only do the seasonal changes create variety, but there is a great variation from one year to another.

Birds often reflect yearly variations very noticeably which makes bird watching quite interesting. One of the birds that is obvious one year and not the next is the varied thrush, *Ixoreus naevius.* Many people ask me about a strange looking robin they have seen in their yard, and surprisingly enough that is a well-used description. When the famous naturalist John Burroughs saw the varied thrush for the first time in 1899 he recorded his thoughts in verse:

"O varied thrush! O robin strange!
Behold my mute surprise.
Thy form and flight I long have known,
But not this new disguise."

The disguise of this member of the thrush family is very distinct. His back is the same slate color as his close relative the robin. The orange eye-stripe, orange wing-bars and the black band across the orange breast make identification easy and positive. The female's breast band is grayish, while the young's breast band is speckled.

The varied thrush or Oregon robin, as it is sometimes called, arrives in our area about the middle of October. They are usually gone by the end of March.

Their summer home is from the northwest part of California north through Alaska. They live in moist coniferous woodlands of the Pacific Northwest. Their call is a long musical note with an artificial sound. As a naturalist at Prairie Creek Redwoods State Park many years ago I was told by a youngster that their call was the water pump squeaking. When I asked the little girl how she knew, she said her daddy told her. What do you do then?

Apples hanging on the tree are an irresistible lure for the varied thrush. They have food habits similar to a robin, and like man's fruit to the point of sometimes being a nuisance.

Watch for the varied thrush. You may not see him every year.

Brown Towhee

Most of the organisms that catch our attention when we look through nature's window are obvious either by appearance or by action. There is one bird, however, that is about as nondescript as a bird can be and does little to single himself out by his actions. I am referring to a member of the sparrow family called the brown towhee, *Piplo fuscus.*

This rather common resident of our valley is often overlooked because of his quite gentle manners and his very dull coloration. He is a rather stout overgrown sparrow. The overall coloration is a grayish brown with our valley variety being more brown than gray. His moderately long tail is darker than the rest of his body, but the feathers under his tail are bright cinnamon brown. This is about the only colorful area on the whole bird, and he keeps this out of sight except for short glimpses. His throat is tinged with a slightly reddish-brown color, and when viewed closely it shows light black streaks. The sparrow-shaped bill is also dusky.

This almost robin-sized bird is a common resident of neighborhoods where bushes and trees are dense and well-established. He is much larger than other sparrows and lacks any wing or head stripes. His very dullness makes identification very easy, and there is really no other bird around to confuse him with.

His method of scratching leaves out of the way has always fascinated me. In some manner he scratches with both feet at once. I have tried this and now have great respect for the towhee. I almost fell on my face.

If you have these birds around your yard, you will find they have a definite personality. It has even been suggested to me by a prominent bird watcher that Snoopy's friend, Woodstock, is a brown towhee. I can't vouch for that but the brown towhee is as close as any I am sure.

Watch for this bird with a jerky tail and pleasant personality. See if yours have character like mine do.

Rufous-sided Towhee

One of the most fulfilling rewards of bird watching is the exuberance shown by those just beginning the pastime. Everytime a new bird is seen, the neophyte is convinced it must be rare because he has never seen it before. Actually, he has probably seen it many times but never realized it was any different. He probably never really looked at it.

The rufous-sided towhee, *Pipilo erythrophthalmus,* is just such a bird. This striking, rather quiet member of our bird community is far more common than most people realize.

The colorful male has a completely black hood that contrasts sharply with his white belly. His back, wings and tail are also black although they all have flecks, spots, or bars of white. The sides and flanks are a robin-breast red called rufous. This gives the bird his name. A fiery red eye makes final identification of this near-robin-sized bird unmistakable. The female markings are quite similar to the male, but the black parts of the male are a dusky brown on the female.

The rufous-sided towhee, sometimes called spotted towhee, is a bird of the underbrush like chaparral, river-bottom thickets, berry tangles, brush patches, and the like. The more dense it is the better they seem to like it, especially for nesting. Nests are built in heavy growth on the ground or in low bushes. The young leave the nest before they can fly but are well protected by their coloration.

This member of the sparrow family eats beetles, ants, caterpillars, moths, crickets, bugs and flies as well as seeds of star thistle, oats, wheat, miner's lettuce, redmaids, poison oak and many others.

They find most of their food by picking through the leaves on the ground. They do this by pulling and kicking the leaves backward using both their feet at the same time—something like a child jumping rope. They do it very easily. I've tried and can't do it at all.

Although they are generally shy, they are common enough that we see them often in the right habitat. They will often perch on an outer branch of a tree or shrub and put on quite a display.

Watch for this handsome character. You will find it fun to be able to name him.

White and Golden-crowned Sparrows

One of the greatest surprises to one who peeks through nature's window is the way an apparently barren tree comes to life when one looks closely. When you look only at the forest you generally see only the forest. When you look only at the trees you may well miss the more fascinating scenes played on nature's stage. Within the branches of a tree an interesting response to the fall season takes place. Without a sound, bouncy new visitors appear in the trees, yards and gardens of our homes. They spend their waking hours in an intense search for the necessary energy of life. Their nervous flitting makes observation a bit difficult for the uninitiated, but with desire and perseverance many birds can be easily identified.

Two of the more common winter visitors are the white-crowned, *Zonotrichia leucophrys,* and the golden-crowned sparrows,

White-crowned Sparrow

Golden-crowned Sparrow

Zonotrichia atricapilla. They both come to us from the North and generally arrive the middle of September. They are both about the same size and shape, and are often found in mixed flocks. The white-crowned sparrow adult has a perky crown with black and white stripes. They are handsome sleek little fellows with grayish breasts and pinkish bills. The younger birds do not have the black and white crowns but have instead a brown and buff colored crown.

The golden-crowned sparrow is quite similar to the white-crown but has no white head stripes. His crown is a wide dull yellow or golden central stripe bordered in black. The crown of young birds usually has no black and the gold is more dull.

Both adults and young can be enticed to your yard by putting out the simplest of foods. Wild bird seed, bread crumbs or corn meal placed on an elevated platform or even on the lawn will generally bring many of these sassy little moochers in to close range so they can easily be observed. Try it. We have only from September to the middle of May to enjoy them.

Oregon Junco

Bobbing bits of black and flashing slivers of white are about as good a description as any for the tiny Oregon junco, *Junco oreganus.* This small bird typifies so very well the fascinating discovery aspect of nature. One can look at nature and miss so much until one learns to really see. I have known so many people that were amazed to see a field come live with birds when they looked at it closely. The Oregon junco is one of the birds that seem to hide so well.

This member of the finch family has a head and breast that are black or grayish-black like an executioner's hood. The rest of the body is white below, rusty above, and buffy on the sides. The tail is black with white outer feathers. The constant flitting of the tail makes this white quite obvious while on the ground or in flight. The rather heavy pinkish-white bill stands out sharply against the dark head. The whole bird is only five to six inches in length—smaller than most sparrows.

Juncos appear in our valley in October and stay with us through April. Most of ours come to us from above 3,000 feet in the Sierra Nevada and are a subspecies that could actually be called the Sierra junco. They are often called "snow birds" because many stay above the snowline all winter. They scratch for seeds in the snow and burrow in the snow for shelter at night.

In our valley they travel and forage in flocks that grow larger as more birds arrive from the mountains. Weed seeds make up the largest percentage of food, and they are frequent visitors to home feeders.

Watch for these almost invisible little visitors. You will enjoy them I know.

Plain Titmouse

One of the joys of having an interest in nature is the intriguing way her secrets are unfolded. When one first starts to observe, only the obvious is seen. As one becomes more attuned, an ever-increasing crescendo of delights is available.

When one first begins to actively look at birds the jays, crows, robins, etc., are the only ones noticed, but soon other less obvious birds come into the scene. One of these generally unnoticed but common birds is the plain titmouse, *Parus inornatus.*

As its name implies, it is really plain. It would seem inconsistent to say that it is easy to identify this cocky little character, but he is easy to distinguish. The plain titmouse is the only small gray-backed bird with a pointed crest. The male and female are almost identical and are found together all year. They are almost six inches long and are gray above and a little lighter below. The sturdy multipurpose bill is well adapted to the finding of bugs, leaf hoppers, moths, weevils, scale insects and insect pupae. The bill is equally suited to crack and eat seeds of wild oats, weeds, melons, squash and sunflowers. Various fruits and acorns are also eaten.

In our valley, we generally associate the plain titmouse with our oaks as they provide them with not only a smorgasbord of insects, seeds, and acorns, but with the crevices, holes and rotted wood that they need to make nests as well.

Normally, the female lays six to eight eggs and incubates them herself. Both the mother and father feed the ravenous brood, however, and then send them out to fend for themselves as soon as the youngsters are able.

Watch for these friendly little feathered neighbors. They are almost always in pairs.

Audubon's Warbler

In winter, the skies become a leaden gray color and the chilling rains make the soil soggy. The out-of-doors becomes a drab, dreary amphitheater for the creatures of nature. It is into this setting that a tiny gray bird with barely a splash of gold bounces quietly. He seems to fit in well with the somber ashen skies till the wings are spread and a bright yellow spot on his rump appears.

This cheery little fellow is called the Audubon's warbler, or now officially yellow-rumped warbler. He is slightly smaller than the tiny Oregon junco. In winter the warbler loses the blue-gray coloring on his back, and becomes a dull brownish-gray.

When you know the field marks of this tiny hank of feathers it is very easy to recognize him from the many other tiny gray birds. His yellow rump spot and throat along with a broad white wingbar eliminates any possibility of confusion with other birds. The tail is twitched frequently, showing many white spots.

It is near the first of September when this visitor arrives to spend the winter with us. While here they eat small fruits, catch insects, and will visit your back yard regularly if suet is placed outside. Suet is the hard fat found on beef and can usually be obtained for a few cents a pound from the meat counter. Bacon grease should not be used as bird food because of the salt content. Most birds physiologically cannot accommodate concentrations of salt.

Before the middle of May is past all of the Audubon's warblers have returned to the Sierra Nevada. There at elevations above 3,000 feet they build a feather-lined nest in a conifer tree. Nesting birds are common throughout the conifer forests as high as 10,000 feet.

If you look through nature's window and see tiny indistinguishable birds that all look alike, keep watching. It will please you greatly to see the yellow rump spot and be able to name this bird.

Ruby-crowned Kinglet

If one takes the time to look through nature's window, he often sees only the very obvious participants in nature's play. He may miss entirely some of the fascinating background players if he is not careful. Even if he is not practiced in the art of observing nature, he will thrill to the majestic soaring of hawks and cannot help but notice the raucous call of the jay. He will often miss entirely the tiny little inconspicuous birds that flit through the trees. The tiny ones, however, can be very interesting and rewarding.

One such winter visitor is a trim, handsome little fellow about one-third the size of the Oregon junco. He has the name ruby-crowned kinglet because of the brilliant ruby red head patch the male can display when aroused. Otherwise, the olive-green color of his back and the lighter underparts tend to make this fascinating backyard visitor almost invisible.

When he finally is seen, however, he can be positively identified by a broken nearly-white eye ring that makes the eye seem very large and dark. There are usually two whitish wing bars present; the rear one is bordered behind by a dark band. The tail is rather stubby and slightly forked.

The behavior of the ruby-crowned kinglet is also quite distinctive. If his nervousness were exhibited by a human, it would typify the extreme antics of a young boy showing off at his birthday party. This tiny creature

keeps constantly on the move in the trees and bushes although he will often come quite close.

The kinglet comes to us in the middle of September from the Sierra Lodgepole-Fir belt above 4,000 feet. He stays in the valley until the middle of April and forages for insects in the trees and on the ground. When he returns to the Sierra Nevada, a bulky nest of bark, feathers and moss is built on a branch of a conifer tree. The female, who lacks the ruby crown, then lays six to nine spotted eggs.

The kinglet can be attracted to the backyard by putting out suet. I am sure you will find the effort well worthwhile. You will be very pleased with yourself when you first see and identify this active little bundle of energy. Watch for him.

Common Bushtit

If one could substitute mammals for birds, the mouse of the trees would probably be the bouncy little bushtit.

The common bushtit, *Psaltriparous minimus,* is the species we have in this area. As the tiniest member of the Paridae family, they weigh less than a quarter of an ounce each and are only four inches long. Half of that length is a long slender tail. They are in the same family as the titmouse and chickadee, and are usually even more acrobatic. Their brownish-gray color is almost uniform although the head often looks more brown to me. In the fall and winter they travel in flocks of 15 or more and seem to enjoy each other's company. They constantly communicate with gentle high-pitched twittering notes.

It is fascinating to see bushtits move into a tree. Usually, you get a glimpse of just one and strain your eyes to see it. Then you realize the tree is full of them. They stay in flocks till late winter and then pair up.

Bushtits build quite an elaborate and ornate nest that may take six weeks to complete. Using spider webs to fashion a horizontal rim, they work downward weaving a 7 to 10-inch gourd-shaped nest of spider webs, lichens, mosses, rootlets and other similar material. Flower petals, oak leaves, blossoms and, sometimes, paper or string are used to decorate the outside. The inside is carefully lined with feathers, down, and hair.

Five to seven eggs not much larger than

beans are laid, and the young grow quickly. Two broods a year are not uncommon. Insects, spiders, plant lice, and scale insects provide the basis of their food supply. They destroy the immense numbers of pests and serve us well.

Cedar Waxwing

Before summer has passed one of the most beautiful of all birds returns to our valley. To me, the cedar waxwing, *Bombycilla cedrorum* is one of the greatest surprises for one who looks through nature's window. From a distance he would not take any beauty prizes because he appears as a dull grayish-brown bobbing bump. This undistinguished little fellow takes on his beautiful appearance when he is observed in detail, either closely or with binoculars. Then you see his glossy cedar brown plumage and his jaunty crest.

His rich brown body is decorated in extremely fine taste. The ends of his tail feathers are a clear brilliant yellow, and often the secondary feathers on his wings have the brightest red tips you have ever seen. These are not the gaudy colors of a parrot but a delicate arrangement of hues

reflecting the work of a gifted artist. Artists, in fact, have often used the waxwing as parts of designs and ornamentation in their work.

The cedar waxwing does not stay here year round but is with us from September to June. In June, the compact flocks depart for the northwestern part of California and parts north for nesting. While these well-coordinated flocks are here they spend most of their time searching out berries. One of their favorites is the pyracantha bush when the orange or red waxy berries are ripe. If one watches, he will see flocks of the birds perch on TV antennas, leafless trees or on telephone lines. They will then singly or in small groups descend upon the bushes and proceed to gobble berries while swinging merrily on the ends of the branches like a happy acrobat.

Sometimes the berries spoil, and products that produce an alcoholic effect in the waxwings are present. If too many of these spoiled berries are eaten, the birds lose coordination and cannot fly. They flop around on the ground and are easy prey for cats. When the berries are eaten along highways the birds are often killed by cars. If you find birds in this condition, they should be picked up and put in a box and kept warm. When they return to normal they can be released and continue to play their role in nature's scheme.

Acorn Woodpecker

One of the common birds seen in our valley is the acorn woodpecker, *Melanerpes formicivorus.* This noisy member of the bird world is about the size of a blackbird. The striking pattern of color is essentially black and white with some red and yellow. The back is a shiny black but with a large white rump spot. The black wings have conspicuous white patches on both top and bottom. The head is colored like a painted clown with a red crown, black chin and patch through the eye, whitish cheeks and throat. The male has a white forehead while the female's red crown is preceded by a black band.

This woodpecker has been known by several other names. In other parts of the country it is often called the ant-eating woodpecker indicated by the formicivorus of

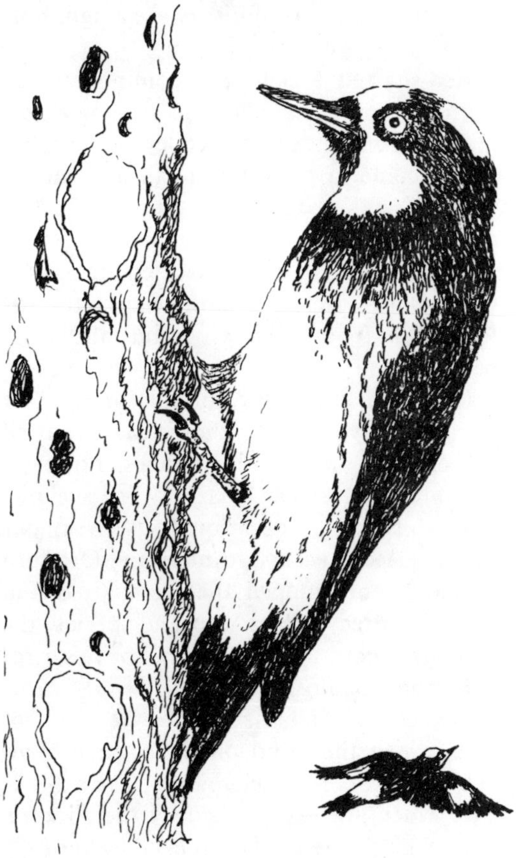

the species name. I have never seen ours eat ants. In California it is sometimes called the California woodpecker.

Acorn woodpeckers are very sociable, and five or six on the same limb are not unusual. When one lands on a limb where others are, there is always a lot of yaking, bowing and wing fluttering. Single birds have even been known to help paired adults feed their young.

Acorns are the main food of this bird as its name would imply. Holes are drilled in soft wood and an acorn is pushed in point first. Cracks or splits are also used and a rotten limb may have hundreds of acorns stored in it. It was estimated that one tree had 50,000 acorns embedded in it. When the woodpecker is hungry he drills out the end and eats the acorn. Undoubtedly, some of the acorns have worms in them and the woodpeckers eat them too. It has been said that the woodpeckers pick wormy acorns, but I have never read any research that proved the point.

Sometimes they store their acorns in cedar shingles or soft beams of houses and make a nuisance of themselves. I suspect this happens when there are no natural places to store their acorns. The oaks are left, but the rotted limbs are pruned out. What else can the birds do?

Red-shafted Flicker

Many of the creatures of nature are difficult to identify, but there are a few that leave no doubt in your mind as to what they are. The red-shafted flicker, *Colaptes cafer,* is one. This woodpecker has so many identifying characteristics that recognizing him is quite easy.

As he flies away, a large white rump spot is quite obvious. He is larger than a robin and has a brown head. His back is also brown but barred with black. There is a large black crescent across the breast that separates the neck from the black spotted belly. This attractive bird gets its name from the

feather shafts which are a salmon red as are the linings of the wings and tail.

The male has red mustache marks down the side of the throat. The marks are also there on the female, but they are a light buff color.

Red-shafted flickers are common in the western United States except for the desert regions. Some stay in our valley all year, but others come in from the north and mountains in winter. They are sometimes called red hammers.

People from the eastern United States may recognize all the field marks except the red feather shafts and red wing and tail linings because the eastern species has yellow feather shafts and wing and tail linings. It is appropriately called the yellow-shafted flicker. The male of this species, however, has black mustache marks. The yellow-shafted flicker is occasionally seen in our valley. Recently the American Ornithological Union lumped the red-shafted and yellow-shafted flickers into one species, the common flicker. It is still fun to watch for variations in color.

Flickers eat fallen seeds, berries and insects which they find on the ground. Evidently the largest single part of their diet is ants which are easily taken by the flicker's long sticky tongue. The flicker is the only woodpecker that regularly feeds on the ground.

Many of the California Indian groups used the red shafts of the flicker in ornamenting their headdresses and other regalia.

Watch for this common ecological partner. He will always be looking out for you.

Sapsucker

When the crisp fall winds wrench the leaves from the deciduous trees of our valley, the handiwork of the sapsucker becomes evident. I am often asked what makes the rows of holes around the limbs and trunks of various trees. The birds responsible are known as sapsuckers. There are only two species found in California; the Williamson's, *Sphyrapicus thyroideus,* is very rare. The red-breasted and yellow-bellied are color variations but are the same species, *Sphyrapicus varius.*

The red-breasted variety is the closest bird in appearance to the eastern red-headed woodpecker, *Melanerpes erythrocephalus.* The red-breasted sapsucker has the entire head and breast a very bright red. The yellow-bellied sapsucker has a red forehead patch with black and white markings on the side of the head. The male has a red throat, while the female has a white one. Both color phases have a long white wing patch running lengthwise. This makes a good field mark. The birds are 6 to 9 inches in length.

The red-breasted sapsucker is the common one here. It breeds in the high mountains and winters in the lowlands and along the Pacific Coast. Both the male and female participate in excavating the tree hole in which 4 to 6 white eggs are laid.

Sapsuckers feed by drilling holes about an inch apart through the bark of trees into the sap. Trees like walnut, apple, cherry, dogwood and cedar are used. The holes are visited periodically and any insects that are attracted are eaten. Ants, beetles, caterpillars, spiders and the like are all utilized as food. The sap in the hole is scraped out by the brushlike tongue. The sapsucker's tongue is different than other woodpeckers in that it does not have the sharp points nor can it be extended as far.

The fact that the holes this bird drills are often in nice neat rows fascinates many. They wonder why the bird does it that way. A few minutes of watching the bird work, however, would answer the question. After drilling one row, the bird merely puts his toes on the edge of the row just drilled and drills the next one up. If too numerous, the birds may harm a tree by girdling it. I have never seen it however.

Watch for this attractive bird and keep an eye out for his handiwork.

Water Ouzel or Dipper

One of the most unique creatures in nature is the water ouzel or dipper.

This drab but fascinating little bird has selected water as his home but has not the webbed feet of the duck nor the long legs of the wader to assist him in his quest for survival.

Instead, he resembles a large slaty-gray

wren with a short wrenlike tail. His chunky body is near the size of a robin.

The dipper, *Cenclus mexicanus,* has chosen not the water of the ocean, marsh, or lake for his home but makes his way where mountain streams and rivers cascade through tortuous self-cut canyons.

One can always expect the dipper where crystal-clear water ripples over ledges, rocks and logs. In fact, his flight is almost never any other place than close to the water of winding mountain waterways. The nest of the dipper is also near rushing mountain waters. Sometimes it is behind a waterfall curtain, on a spray-covered rock or under a secluded overhang of the bank.

When searching for food, this "songbird of the streams" displays his most unique adaptation. Although he may wade in shallow water, he does not hesitate to plunge into deeper water and actually walk below the surface.

His natural buoyancy is overcome by partially extended wings that flap in a complicated manner to keep him submerged.

To complement this technique, he has a movable flap over the nostrils to seal out water, a highly developed nictitating membrane to protect the eye, soft plumage and thick down for insulation and a preen gland for waterproofing ten times the size of any other small passerine bird.

The most obvious characteristic, however, is his bobbing or dipping which is a frequent, rapid, bending of the knees. When a bird is seen doing this near a mountain stream, it almost has to be a dipper. Watch for him in your visits to the mountains.

Turkey

At Thanksgiving everyone has the turkey on his mind. No other animal is as closely associated with a holiday as the turkey is with Thanksgiving in the United States. Therefore, I think it is appropriate to know a little more about it as a bird instead of just as the centerpiece on the Thanksgiving dinner table.

It is thought by some that originally the turkey was mistaken for a guinea fowl species believed to have come from Turkey—a general name for all Moslem countries. Others believe the name came from one of the turkey's calls—turc, turc, turc. Still others contend the name is a corruption of the Indian name for the bird "firkee."

As a bird, the turkey had some notable admirers, not the least of which was Ben Franklin. He formally objected to the selection of the bald eagle as our national emblem and indicated his preference as the turkey. He felt the turkey and its characteristics better represented the United States.

The turkey, *Meleagris gallopavo,* is a completely American bird although our domestic turkey came to us from Europe. This sounds inconsistent but is a matter of fact. In the sixteenth century the Spanish conquerors invaded Mexico and found the natives raising a large toothsome bird. This bird, the turkey, was taken to Spain where it ultimately was introduced into France and England. Emigrants then brought this Mexican variety, *Meleagris mexicano,* to the United States. The early colonists, instead of shooting wild birds, probably had domestic birds for their Thanksgiving feast. The Mexican variety then freely bred with our North American varieties, making pure strains of our native wild turkey rare.

The wild strains have chestnut instead of white tips on the tail feathers and upper tail coverts. They are sometimes referred to as a streamlined version of the barnyard turkey.

Their food is largely seeds, nuts, grain and insects. They can fly with great speed.

The male wild turkey typically weighs about 20 pounds while the hen weighs only 10 pounds. The male may have a harem of up to 15 hens for which he gobbles and fights. He plays no part in nesting, and it has been mentioned he may break the eggs if he finds a nest.

This magnificent bird is described as the grandest of all America's game birds. Too bad they are so scarce in the wild.

6
Plants

Lichens, Fungus and Grasses
Trees
Shrubs
Wild Flowers
Useful and Edible Plants
"Weeds"

Some of the flowers that nature uses are often of great singular beauty, but more often the dazzling displays of brilliant color are created by rather ordinary blooms.

Old-Man's Beard

In the world of nature there are many items of interest that are frequently misunderstood or misidentified. One that is constantly brought to my attention is the pendulus tree lichen called Old-Man's Beard, *Usnea barbata.* This lichen is quite often erroneously called Spanish moss because of its similarity to the true Spanish moss found in the southern part of the United States. The Spanish moss, *tillandsia,* is a flowering plant that is related to the pineapple plant in the Bromelia family. It is not a moss at all, but it does use trees to grow on. It makes its own food and gets its water from the air. In this, it is like our own old-man's beard.

As I mentioned, old-man's beard is a lichen. Lichens are actually two entirely different kinds of plants that join together for mutual benefit. The matrix of the plant is a fungus which absorbs and stores water while the food making is performed by algae cells embedded in the fungus. Tiny threadlike parts of the plant called hyphae produce strong acids that can even decompose rock for their minerals. In general, lichens are highly resistent to cold, drought and strong sunlight but very susceptible to smog or air pollution. Lichens can and do grow on frigid mountain tops, the arctic region and on bare rock without soil. They are often the first plants to grow in an area and are often the pioneer plants.

Old-man's beard, a fruiticose lichen, is made up of numerous fibrous branches that are usually some shade of green. It is seen hanging from branches or bark of living or dead trees throughout most of North America. The fibers are often more than four inches long.

It was described as early as 300 B.C. and the power to grow hair was attributed to it. Too bad it doesn't work. It could make a lot of people happy. The ecological contributions, however, are very important as they are often the first step in soil building.

If you see any long, stringy, fibrous plants growing from branches of trees in our valley, it is probably old-man's beard. It is not Spanish moss.

Mushrooms

As the summer breezes give way to the winds of fall the climate of our valley changes drastically. The long drought of summer comes to a close when the first fall rains settle to earth and penetrate the sun-baked soil. These rains produce many obvious changes in what we see as we look through nature's window.

It is easy for one to notice the warm brown hills change to a lush green as the new moisture spurs the transformation of dormant seeds into plants. One may not realize the almost phenomenal appearance of various members of the fungi world unless he is constantly searching nature's face.

It is not unusual to have a beautiful lawn composed of a uniform crop of grass utilizing the fleeting fall sun one day, and the next day to have numerous creamy-brown mushrooms rising far above their chlorophyllic neighbors.

This sight quite often creates apprehension in the minds of mothers with young children. The tendency of children to put things in their mouths is well-known, and the toxicity of toadstools is well publicized. The combination of children and toadstools is always a natural potential disaster.

Perhaps the main reason there are not more casualties from toadstools is that almost all of the fungi found growing in our yards are harmless. This, however, is little solace to a careful mother. She should always check the yard before letting her children play and if mushrooms are found, they should be removed.

Notice I used the term mushroom in referring to the fungi illustrated. Many people believe that the edible ones are mushrooms and the poisonous ones are toadstools, but such is not the case. They, the kind with caps, are *all* mushrooms. The poisonous mushrooms are called toadstools.

Shaggymanes

Now and then we read in the news about people who have eaten poisonous species of mushrooms called toadstools. The inherent danger of eating mushrooms cannot be minimized, and the identification of wild mushrooms must be absolutely positive before eating.

There is no single safe rule that can be used to distinguish poisonous from non-poisonous mushrooms. There are, however, a few edible wild mushrooms that can easily be positively identified by sight. Once they are learned, they can be gathered and eaten safely.

One of these mushrooms is the shaggymane, *Coprinus comatus.* This large, meaty mushroom is one of the best of the common edible kinds. The characteristic cap is nearly cylindrical when young and from 1 1/2 to 2 inches wide. The cap grows four to six inches high and spreads at the base as it gets older. The name, shaggymane, refers to the soft shaggy scales that are found on the cap.

The white stem is about one-half inch thick and five to eight inches long. It is also rather fibrous and brittle, and the ones I have found have been hollow.

The shaggymane is common throughout the country and is found growing in grassy areas like parks, lawns, and golf courses. They usually grow in groups and are sometimes tightly clumped together. They are most common in the spring and fall. My class and I found them growing in Del Paso Park and had a very tasty snack.

Unlike many other kinds of mushrooms the shaggymane cannot be stored, even in the refrigerator. If it is stored, it slowly liquifies into a black fluid. Often the cap is black at the edges from the same process. This black should be cut off before eating the cap. Before eating any mushroom, it should be examined for maggots or other insects inside. Cut lengthwise through the cap and stem to check.

Mushrooms can be enjoyed if the simple rule of positive identification is followed.

Bermuda Grass

To the home owner in our valley there is no plant any better known than Bermuda grass, *Cynodon dactylon.* This plant is obvious enough in the summer, but is generally even more obvious in winter. I know this seems peculiar since few plants are showy in death. Bermuda grass, however, dies and turns brown in winter, and the mats it forms in spring and summer become very thick and tough. Bermuda lawns, although hardy and hard to damage, turn brown in winter and are generally considered unsightly.

Bermuda grass is a perennial that produces stems that may be underground or on the surface. The rootstocks may be on the surface as well as underground. Both rootstock and runners can send up leaves, stems, and roots from each joint. These may grow out many feet from the parent plant. The blade of the leaves are about one-eighth of an inch broad and one to four inches long. The somewhat similar plant, crabgrass, has leaves some of which are almost one-half inch wide. A very distinctive characteristic of Bermuda grass is the ring of hairs on the outside of the sheath of each leaf.

Bermuda grass is a native of the warmer parts of Europe. It was named after the Bermuda Islands where it is quite common. It is also called devil, scotch, dog's tooth, and wire grass. It is well established in the southern part of the United States and the warmer parts of California. It is now one of our most troublesome weeds. It chokes gardens, covers road and walk edges, and crowds out more desirable lawn grasses. It is spread by seeds as an impurity in crop seeds, by parts of the grass carried on farm equipment, animal bedding, wind and countless other ways. It is very drought resistant and tolerant of alkali. It does not flourish in dense shade and can sometimes be controlled by a smothering cover crop. The roots when exposed to the sun die quickly. In areas where there are prolonged freezing temperatures Bermuda grass does not establish itself.

There is a lot of this grass in our valley. Watch for it and see if you can tell it from crabgrass.

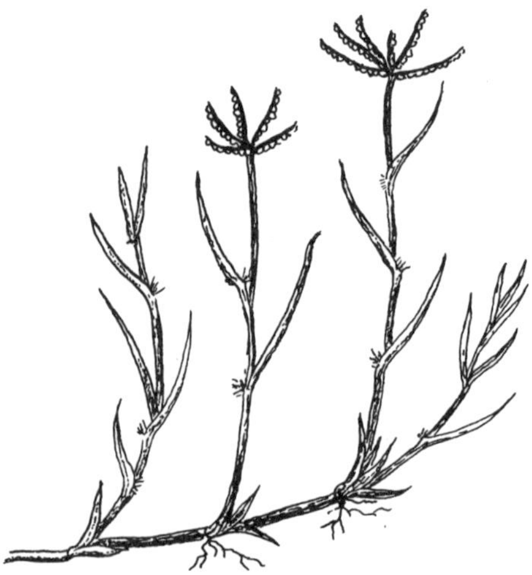

Dallis Grass

Grasses are the commonest of flowering plants and serve many roles in the complexities of nature. They are the staple food for ground squirrels, marmots, meadow mice, rabbits, grasshoppers and many other insects. Alive or dead, grasses and their roots serve as soil binders to lessen erosion by wind or water. They turn our valley green in spring and golden brown in summer.

There are dozens of species that vary in structure, growth form, and distribution in our valley. Dallis grass, *Paspalum dilatatum,* is one of the most showy and easily identified grasses when it blooms and produces seeds in late summer.

Dallis grass is a perennial that grows from a short rhizome. The leaves are not quite one-half inch wide and are up to 10 inches long. There are usually 3 to 5 racemes or spikelets on a culm or stem. The racemes can be up to 8 inches long and usually point upward and spread out. The seeds are in crowded spikelets and are less than a quarter of an inch long. Silky black hairs protrude from the seeds.

Dallis grass resembles knotgrass, *Paspalum distichum,* in many ways. Knotgrass, however, normally only has 2 paired spikelets while dallis grass has 3 to 5.

This grass was introduced from Uruguay and Argentina as a valuable pasture grass. It grows in a variety of habitats like dry prairies, marshy meadows, borders of ditches and canals, and roadsides. It also grows in my yard where, I assure you, it is not wanted, especially by my sons who mow the grass.

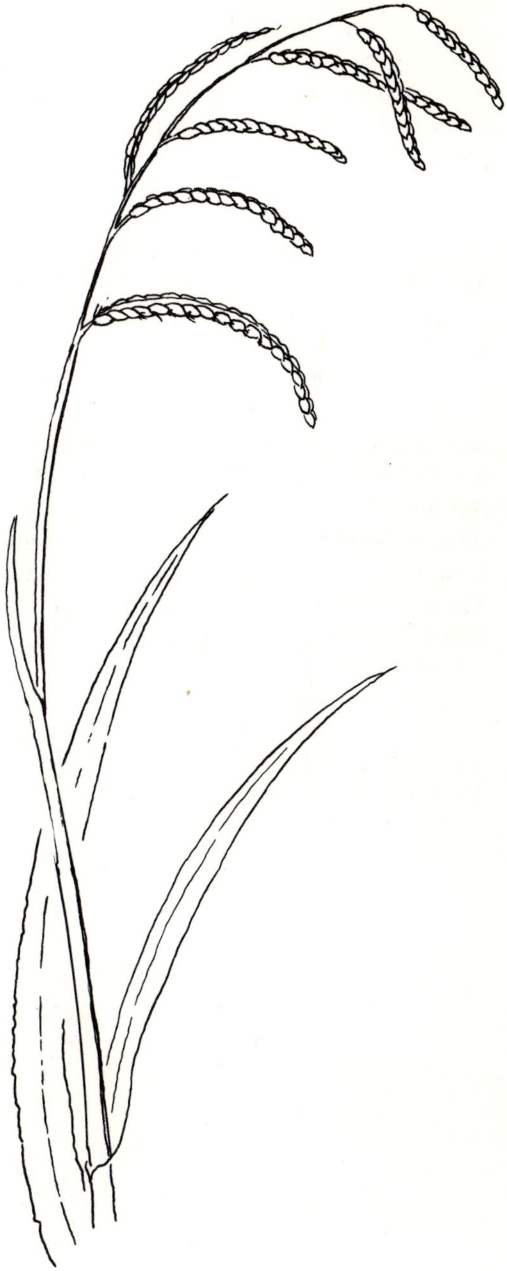

Cattails and Tules

One of the interesting facets of our environment is the variety it offers those who look through nature's window. One need never be bored with the smorgasbord nature presents. Not only does our environment vary from one locale to another but also within the locale. There are many differences in grasslands, woodlands, brushlands and wetlands, and when one looks closely, various plants and animals are characteristic of each habitat.

Two of the most commonly misidentified plants are found in a wetland habitat. I am referring to the widely distributed cattail, *Typha latifolia,* and the common tule, *Scirpus acutus.* Cattails and tules occupy much the same kind of swampy or moist areas and as such are often mistaken for one another. There are, however, major differences that make identification quite simple.

Cattails have flat grasslike leaves that may grow more than 6 feet tall and be up to one inch broad. The spike that is a familiar sight this time of year is actually a flower cluster. Located on a round stem, the 7 to 13-inch spike is composed of two parts: the pollen-producing part or male flowers above and the pistillate or female flowers below. These are often gathered for decorations and when kept too long may give the housekeeper quite a surprise. When the spike reaches a certain ripeness, it releases the individual seeds clothed in its cotton. The volume and numbers of these are overwhelming.

The leaves of the tule are almost inconspicuous, and the only visible part of the plant is the somewhat hollow stem and its clusters of tiny flowers at its tip. They may grow up to 9 feet tall and are found in saltwater as well as fresh water marshes. Little else need be said about identifying these two plants. Next time you are in their habitat call them by their correct name. It makes the world of nature more fun. Try it.

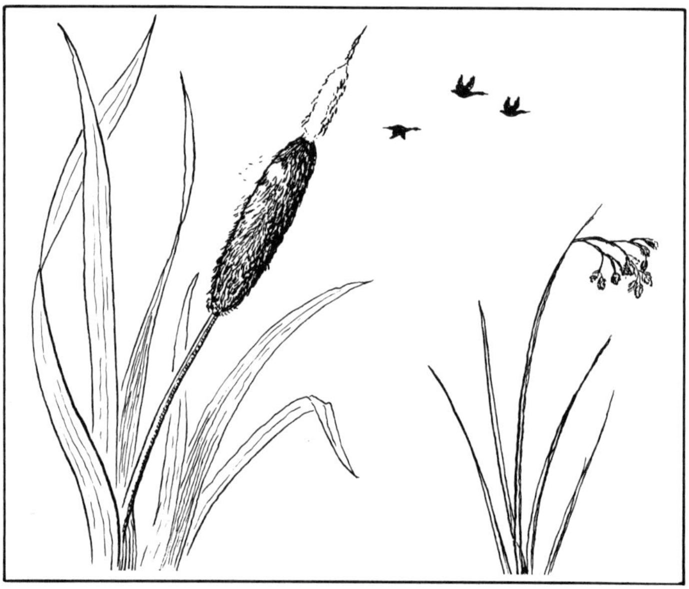

Buckeye

One of the most obvious changes that is evident in the fall is the transition many plants make. To me, the drying of the grasses to a golden brown in late summer transforms the rather monotonous contours of the foothills into a colorful and evident representation of our Golden State. As you travel through our valley the warm brown plains create an aura of tranquility that artists constantly attempt to capture.

Although most of the changes in the hue of our vegetation is expected and accepted, there is one plant that makes a dramatic change that causes concern in the discerning observer.

Every year near the end of summer many people are alarmed at the buckeye tree that has all the symptoms of dying. These trees are generally seen along the streams of this area. The ones bordering American River College are good examples. Here, there are several trees with the leaves brown and hanging down as one would expect a tree in the throes of death to appear. It is as if each leaf has echoed the despair of the tree in death. If one looks more closely, there are lemon-sized grayish-green balls hanging among the leaves. This would seem to indicate even further that something killed the tree before the fruit has ripened. Often there are many of these trees in the area that all look alike. This is even further evidence indicating a disease has attacked and killed all of the trees.

Such is not the case, however. The trees are not dead or dying but merely preparing for winter. The leaves have finished making food for the tree. The leaves then die to prevent further water loss. The various mechanisms the tree has to bring this about are interesting but complicated. The fruit continues to ripen and the leaves and fruit fall later so the fruit can take advantage of the early spring rains. If the fruit is examined after the covering has split open, the rich red brown fruit of the buckeye tree can be recognized.

Chinese Tree of Heaven

I am sure that almost all of our valley viewers through nature's window travel into the Mother Lode Country at some time or another. Visiting old gold mining settlements is a favorite pastime for many people, and pictures of the old towns like Coloma, Columbia and Volcano adorn many walls. Pictures of these old towns almost all show a very characteristic tree that is not native to California.

The Chinese tree of Heaven, *Ailanthus altissima,* is a tall, hardy tree that may grow 100 feet high. It has large leaves composed of 15 to 30 leaflets as shown in the sketch. Trees are of a single sex and may be either male or female. The male tree, in flower, has an unpleasant odor while the female tree produces winged fruits somewhat like the ash. When the deciduous leaves fall, they leave very conspicuous leaf scars on the branch.

The wood is coarse-grained, weak, soft, pale yellow and not durable. It is possible that contact with the leaves may cause a skin irritation. Ants may be drawn in large numbers to the unpleasant smelling leaves.

The tree, sometimes called Chinese sumac or ailanthus, was introduced by Chinese miners from their homeland. The name ailanthus is from the word *ailanto* meaning "tree tall enough to reach the skies" (Heaven). As I understand it, the tree was planted by the Chinese wherever they went—not as a religious symbol but as a kind of good luck omen. This was certainly appropriate for the gold fields.

The most unique characteristic of the tree is its ability to live and stay green along busy downtown city streets. It seems able to grow even when its roots are completely covered by concrete. It is smoke resistant, is remarkably free of pests and diseases and will even survive occasional inundations of sea water. It is well adapted to live in the polluted world of man. We should be so lucky.

Cottonwood

One of the most attractive features of our valley is the many waterways that thread their way to the Sacramento River. These riparian areas are fascinating and provide us with much of the endless variety that makes living here so pleasant.

If one flies over our valley, the waterways are very obvious by the dense tree growth that marks their path. Valley oak, live oak, blue oak, and Fremont cottonwood are generally the predominant vegetation.

Fremont cottonwood, *Populus fremontii,* may grow to a height of 90 feet. Five-foot thick trunks have been found. They generally have a round-topped crown. The bark is very rough, deeply furrowed and thick. Generally, the bark is a shade of gray but reddish within. The triangular or roundish leaves are smooth, leathery and rather thick. They are edged with roundish teeth and are a yellowish green. Typical leaves are 2 to 4 inches across and are wider than they are long.

As a member of the willow family, the tree produces male and female catkins. The male catkins are not unusual, but the female catkins produce seeds with long cottony hairs attached. These hairs are very obvious in the fall when the wind carries them all around. Many people are allergic to the pollen of the cottonwood which often makes the trees unpopular around houses.

The wood of the Fremont cottonwood is soft and pale brown in color. It is fine-grained, brittle and not durable. It often splits when drying and has little commercial use. Fremont cottonwood is much heavier than other cottonwoods.

Undoubtedly, the primary value of cottonwood is its part in our ecological scheme. As a rapid growing tree, it is an excellent soil holder. Raccoons, wood ducks, etc., use the rotted cavities for homes while beavers, deer, insects, etc., use the foliage and branches for food. The limbs and trunks provide nesting places for a variety of birds.

Cottonwoods were mentioned more often than any other tree in accounts of early exploration. John Fremont mentioned the cottonwoods as "harbingers of better country." Because of his accurate description of the tree it now bears his name.

Digger Pine

One of the most interesting displays in nature is an area where two conditions of life come together and mingle. In these locales the variety in the expression of the environment is fascinating. They also give the observer a glimpse of the most extreme conditions the various species will endure.

Many of our valley residents scarcely realize there is a vegetation change as they travel into the foothills although they may notice that the grasslands of the valley give way to brushlands. There are, however, many other subtle indications of a changing climate as we wind into the Sierra Nevada.

If the traveler looks through nature's window on the way, he will notice a rounded, grayish tree with long needles that is different from the oaks. Although it is a pine tree, it does not have a stately single trunk like most other pines. Many pines seem to convey an aspiring purpose to reach into the heavens. Instead, this tree, known commonly as the digger pine, *Pinus sabiniana,* has a very flowing, relaxed appearance. The upper branches generally grow into u-shaped forks that can be seen from a great distance. The needlelike leaves are from 8 1/2 to 12 inches long and in groups of three. The grayish color gave the tree its earlier name gray pine.

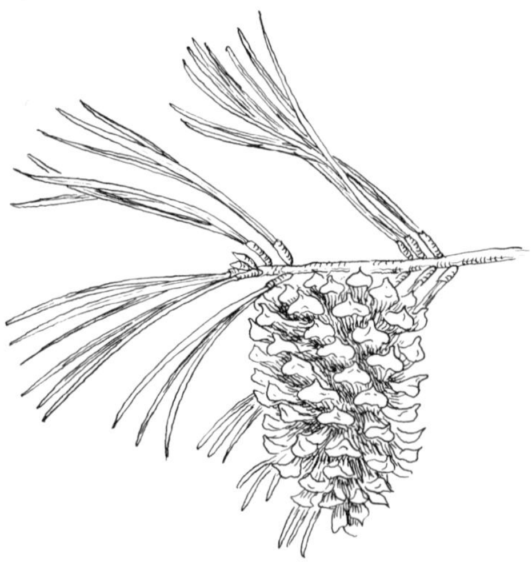

The name digger pine was given because the nuts, found in the large sometimes 12-inch cone, were used by Indians for food. Because the early settlers in California mistakenly called the Indians "diggers" they attached the name to the pine.

In order to gather the seeds the cones were piled and burned. This caused the scales to open and the seeds to fall out so they could be collected. Today, the California ground squirrels harvest most of the nuts.

A few individual digger pines are found on the valley floor bordering the foothills, but the optimum growth occurs between 2,000 and 3,000 feet. They can grow in soils that are gravelly, shallow and dry and can endure rainless summers where even brush is killed by drought.

The digger pine resembles a dusty ghost from a desert scene, but its ghostly qualities do not stop there. The needles are stiff and wiry and stand out at jaunty angles to the branch. This and the sparseness cause the diggers to cast little or no shadow in the sunlight. The next time you see one look for yourself.

Live Oak

A characteristic of fall and winter in our valley is the influx of many feathered visitors from the north and high Sierras. These friendly little fellows are fascinating to watch as they move about through our yards, gardens and fields. It is very convenient that many of our trees have shed their leaves, making our bouncy tree dwellers very obvious. Some of our trees, however, do not lose their leaves.

One of our common native trees that is evergreen is the interior live oak. The oak family is very interesting since some are deciduous and others are evergreen. There are two deciduous oaks that are common in the valley, the blue oak and the valley oak. Both of them are classified as white oaks and have lobed leaves.

The interior live oak is classified as one of the black oaks and is the only evergreen oak common here; hence, the name live oak. There are several other live oaks found in California.

The scientific name of the interior live oak is *Quercus wislizeni.* It is a round-headed tree growing to a height of 75 feet. The leaves are one to two and a half inches long, shiny green above and pale yellowish-green below. The leaves remain on the tree about two years and are always flat, never curled. Their overall shape is oblong to elliptical. The leaf edge is either smooth or spiny toothed. The acorn is a reddish-brown nut about 1 1/2 inches long and takes two seasons to mature. The grayish bark is brittle and thick.

The wood is brittle, hard, heavy and exceedingly fine grained. The heartwood is reddish-brown and the sapwood thick and whitish. It has value only as fuel since the wood cracks badly when seasoned. It is very tolerant of shade and grows in pure or mixed patches on a wide variety of soils. It adapts its growth form to the quality of soil.

This handsome tree is a real asset to any community. The oaks, in general, provide a smorgasbord of epicurean delight for many of nature's creatures. This quality endeared the oaks to the late Effie Yeaw of Carmichael who dedicated much of her very productive life to the planting of trees, especially oaks.

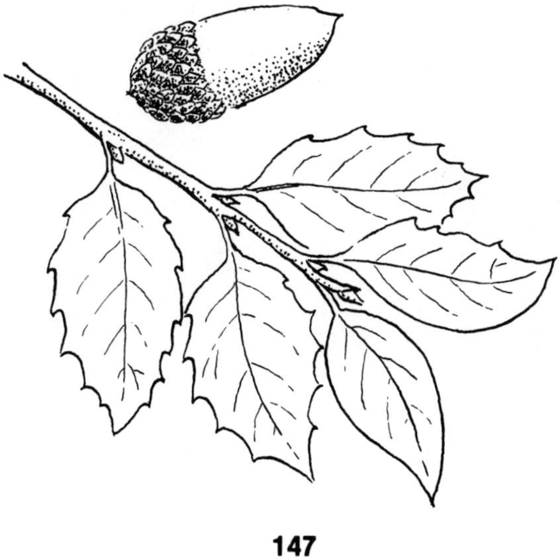

Acorns

"Did you ever eat a pine tree?"

This was a common question asked by a nature food enthusiast, the late Euell Gibbons. To a similar question, "Did you ever eat an oak tree?," most of us would say "No, but I would like to try."

I do not believe the use of acorns for food is going to solve the problem of high prices in the grocery store, but it could be a lot of fun.

The acorn is the fruit of the oak trees and is quite abundant in our valley and foothills. Fall is the time of year to gather acorns if you would like to eat them.

Acorns from all of the oaks were used by the Indians, but they kept the different species separated. The valley oak produced one of their favorite acorns, however, because of the size and sweetness.

The Indians usually soaked the acorns overnight before shelling them. The kernels were then dried.

The dried meats were placed on a large rock. A smaller rock was used to smash and grind them. This process wore holes in the larger rock.

An Indian family might collect and process four or five hundred pounds of acorns each fall.

Acorns cannot be used for food until they are leached and the bitter tannic acid washed out. Indians often did this by making and packing a hollow in the sand. They put the ground meal in it and poured water over the meal several times until the bitterness was gone.

The flat of the hand was then placed on the meal and lifted. The dough that stuck to the hand was scraped off. It was then ready to use for soup, bread, etc.

If you would like to use modern methods to do this, soak and crack the acorns much as the Indians did. Dry them and then grind in a blender. Leaching can be done through a cloth. A good method is using a clean sock tied to a faucet with the acorn meal inside and a trickle of water allowed to run through the meal overnight.

Then bake it, cook it, or whatever. Do not be surprised if it tastes bland. The flavor must be added.

Liquidambar or Sweet Gum

The atmosphere of fall is created by several subtle changes in nature's wardrobe. The crisp morning air and shadowy chills of evening are but a few of the endless variations nature uses to fulfill her intriguing role. I am sure many are sad to see summer come to an end yet look forward to the variety fall offers in her weather. One aspect of fall everyone enjoys, however, is the spectacular hues her broadleaf trees can create.

One of the most colorful trees we see in our valley is the liquidambar or sweet gum. Although there are six species found in the world only one, *Liquidambar styraciflua,* is native to America. It is an important timber tree within its natural range. This range extends from Connecticut to Illinois, south to Florida and into Central and South America. It is one of the most popular ornamental trees in our valley because of the brilliant crimson its leaves turn in fall. These trees are easily identified at this time of year.

The somewhat maplelike leaves have tufts of hairs in the axils or joints of the main veins and may have five to seven lobes. The spiny balls found on the tree produce the seeds and are a nuisance when they fall on the lawn. They make quite a racket when picked up by rotary lawn mowers. They are often used in dry flower arrangements and decorations, often being sprayed a different color.

The names liquidambar and sweet gum refer to the fragrant, balsamlike resin called storax or styrax. This viscid substance has derivatives used in perfumes, soaps, cosmetics, incense, lacquers, medicines and to flavor tobaccos. Our native species is not an important producer of resin, but the wood is very useful for furniture, veneers, boxes, barrels and baskets. It is sold under the name red gum or yellow gum according to its color. It does well in moist, loamy soil and may grow over 100 feet high.

Watch for this outstanding tree and take a closer look at it. If you don't have one in your yard, I'll bet your neighbor does. Crush a leaf and smell it. You may be surprised.

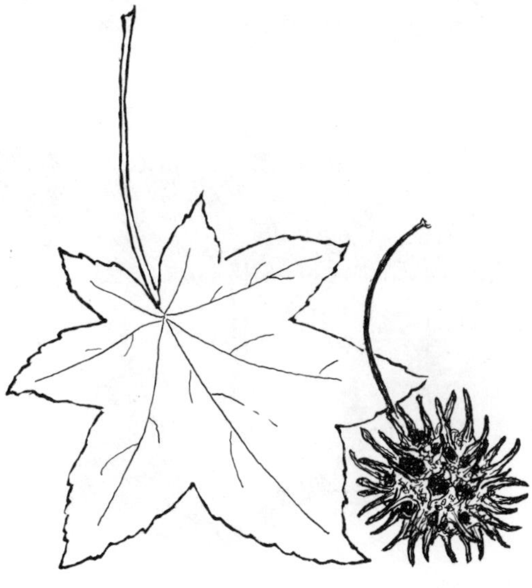

Christmas Trees

The approach of the Christmas season means many preparations. Buying gifts, food, and Christmas cards seem to be a part of getting us in the mood for this very joyous time of year.

For young and old alike, however, the Christmas tree itself is often the center of attention. The type of tree chosen today varies greatly. Some accept clean, convenient, artificial trees, while others go on trips up into the snow and spend great effort to cut just the right one. Most of us visit the Christmas tree lot to select the one for our home, and even though great care is taken in the selection, many people know little or nothing about the tree they buy.

The native trees that we use for Christmas are very easily identified, but, perhaps, a few words on them will be helpful.

A few people like to use pine trees like the sugar pine, *Pinus lambertiana,* or ponderosa pine, *Pinus ponderosa.* Pines have their leaves or needles in bundles, and if you pull them off, they will be in groups. The ponderosa is typically in bundles of three, and sugar pine in bundles of five. The needles of both are more than three inches long.

Incense cedar, *Libocedrus decurrens,* seems to be increasing in popularity. Its flat aromatic branches do not shed in the house, and its fullness makes decoration simple. It is heavy, however, and not traditional enough for most.

One of the more popular and inexpensive trees is the Douglas fir, *Pseudotsuga menziesii.* It is not a fir but does have similar needles and appearance. The needles are blunt or rounded at the tip with a slight groove on top and a ridge below. They are usually 1/2 to 1 1/2 inches long. Although the tree is often full and beautiful, the limbs are not strong enough to take heavy ornamentation and they are not noted for lasting too long.

Perhaps the two most popular trees are white fir, *Abies concolor,* and red fir, *Abies magnifica.* These two trees are often very similar in appearance, and some look much like Douglas fir although the tips of the leaves are usually sharp pointed. The red fir is usually sold as silver tip and the highest prices are paid for them. They are generally cut from areas higher in elevation than the Douglas or white fir, and they are usually longer lasting. Their limbs are able to take heavier ornaments. To tell the white fir from the silver tip look at the base of the needles. If you see a slight twist in the needle, it is white fir.

Some people are concerned that the cutting and buying of Christmas trees is poor conservation. However, when the trees are cut legally and according to conservation practices, no harm is done. In fact, it is often beneficial.

No matter what kind of tree you select, enjoy it. Enjoy the fragrance, the touch of wilderness, and the spirit of Christmas it embodies.

Broom

A pleasant treat is in store for those that do a little traveling through our foothills in the spring. I am referring to the outstanding displays of broom. The brilliant yellow blossoms that cascade over the plant almost obliterate the greenery while the pleasant fragrance reminds one of a giant bouquet.

There are several species of broom found in California. Most have yellow blooms, are six to ten feet high, and have small leaves. The classification into species seems somewhat nebulous and everyone does not agree which is which.

Spanish broom, *Spartium junceum,* is sometimes confused with French, *Cytisus monspessulanus* and Scotch, *Cytisus scoparius* broom, but the branchlets of Spanish broom are rushlike or rounded while the other broom branchlets are very angular.

Brooms are natives of Europe, Spanish broom being of Mediterranean origin. *Spartium* is from the Greek word *sparton* meaning "broom," and branches of the plant were tied together by early people to make brooms. Cytisus is from the Greek word *kutisus* that indicates "a type of clover." Brooms are legumes as are clovers.

Planted as an ornamental and as a soil holder, brooms have become a weed in many areas. Although they are beautiful and fragrant, their dense stands sometimes keep trees from growing and they may even be impenetrable to deer. The vegetation avoided by livestock is poisonous.

An interesting historical fact is told of Geoffry of Anjou. He often wore a branch of broom plant in his cap which led to his being given the surname Plantagenet. *Plante de genete* means "sprig of broom" in French. England was ruled by the Plantagenets from 1154 to 1485 A.D.

The seeds of broom can be roasted and used as a substitute for coffee. The pods and lower buds can be made into good pickles by soaking them overnight in a heavy brine of white wine vinegar. They must be weighted down however. After rinsing they can be cured in any good pickle solution till they do not float. They can then be eaten anytime. The fresh buds should not be eaten as they contain poisonous alkaloids.

Watch for this striking bush. An excellent display can be seen in the spring, on the north side of the Sunrise Boulevard Bridge across the American River.

Coyote Brush

A ride through our surrounding foothills in winter can be very enjoyable. As the rising sun takes the chill out of the morning air it leaves lush green winter grass brilliant with glistening beads of just melted frost.

In winter, it seems unnatural to see a lush green shrub that is sometimes found in great profusion. It does not seem natural to see light-green leaves when most of the other leafy shrubs are grayish or waxy. The most common name used for this different plant is coyote brush, but it is sometimes called chaparral broom. The scientific name is *Baccharis pilularis* variety *consanguinea.*

Baccharis pilularis is a low-growing shrub that forms dense mats under a foot high on the windswept dunes and headlands along our coast. The variety *consanguinea* may grow 15 feet high but is more often around five. It is a very well branched bush. The leaves usually have 5 to 9 teeth and are not quite an inch long. They are less than one-half inch wide.

The most obvious characteristic of this plant is that many of them are covered with a profusion of soft bunches or heads of soft cream-colored hairs. This type of head is common to the sunflower family of which this plant is a member. These hairs protect the seeds and are a part of the fruit. The plant is a very late bloomer, putting its yellowish-white flowers out from August to October. The plant is dioecious which means it has the male and female flowers on separate plants. It grows on the west side of the Sierra Nevada surrounding our valley, but it is seldom found above 2,000 feet.

The names of this plant are a puzzle to me. The name Baccharis is from the Roman and Greek God Bacchus. He was the god of wine. Why that name is attached to this group I don't know. The name coyote brush may have been given to the plant because of the denseness of the hairs on the seed pods resembling the fur of a coyote. That is really stretching things.

Watch for this attractive plant in your travels. I have heard it called the hospitality plant because of the number of insects it supports. Take a close look and see what you can find.

Blue Elderberry

One of the most obvious large shrubs in our valley is blue elderberry, *Sambucus caerulea*. This large shrub may grow to 10 feet and sometimes even 30 feet. It may be considered a small tree. It is rather widespread in the West and is found from the waterways of the valley up to 5,400 feet. The characteristic stalk of the leaf is 5 to 8 inches long and bears 5 to 9 leaflets. These leaflets may be up to 6 inches long. The edges are finely toothed and somewhat smooth.

The white or cream flower clusters are very obvious in spring and early summer. The flat-top clusters may be more than 8 inches across and cannot really be confused with any other shrub. The tiny individual flowers are 5-lobed with 5 stamens. These flower heads can be dipped in a batter and fried quickly making a delicious fritter.

In late summer the flowers ripen into a black berry with a whitish dustylike covering making them appear blue. This berry is about 1/4 inch and very showy. They make some people sick when eaten raw but do not seem to bother others. Many birds and mammals utilize the berries. Band-tailed pigeons, house finches, flickers, mockingbirds, rabbits, squirrels and deer are but a few. When cooked, the berries are edible for all, and elderberry pie is one of my favorites. They can be used fresh, or dried in the sun and then boiled in sugar for making pies later.

The stems of elderberry are filled with a soft pith that can be easily removed. As a boy, I learned to make whistles from them. Indians, however, knew about its musical qualities long ago. They called it "the tree of music." They made flutes and dance clappers from the branches by using a hot stick to burn out the pith. Even the scientific name implies its musical origin since *sambuke* in Greek means "a musical instrument made of elder wood."

Watch for this handsome shrub and its berries. Recipes for pies are easily found; however, care must be taken nowadays when gathering berries along roads. Some counties still spray vegetation, and the berries could possibly be poisonous. Just make sure they are safe, and you are in for a real treat.

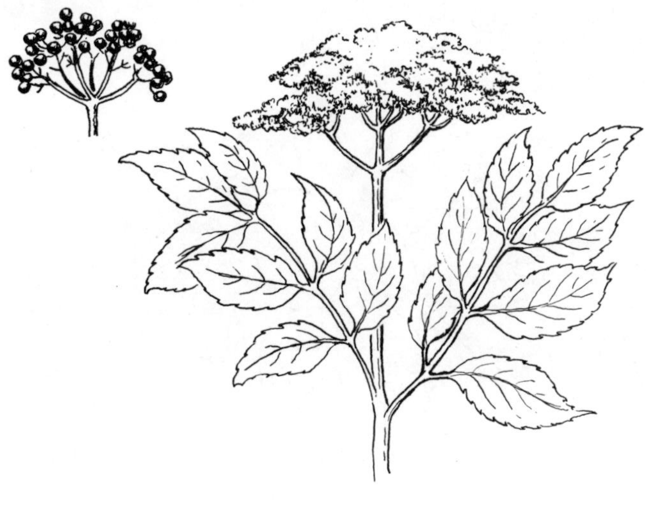

Redbud

A trip through the foothills in late February, March, and April can be a very rewarding experience. For most of the year the chaparral plants do not have much color but in early spring the redbud, *Cercis occidentalis,* changes all that.

Many of the chaparral plants are in bloom then but most have white flowers. The beautiful magenta pink to purple flowers of the redbud stand out strikingly from all the rest.

There are several kinds of redbud throughout the world. Ours may grow to about 20 feet with many stems from the base. The leaves are shed each year and have an attractive heartlike shape. The flowers have 5 petals and appear in clusters before the leaves. The leaves appear later and are bronzy-green at first. They later turn glossy-green.

The fruit is a pod up to 3 inches long that turns a dull red when mature. The seeds have a high rate of germination often up to 70 percent. They will germinate much faster if hot water is poured over the pods. Redbud may be found growing at elevations up to 5,000 feet.

The bark of the redbud stems was sometimes used by Indians to make the designs on their baskets. The stems were also used as basket foundation material. The flowers have a sharp acid flavor and are an asset in salads. The buds, flowers, and young pods are tasty if fried in butter. Indians and early Californians used the bark raw as a medicine for diarrhea.

One of the other kinds of redbuds found in Asia is called *Cercis siliquastrum.* The legends attached to this species gave use to the other common name applied to all redbuds—the Judas tree. One legend states that this is the kind of tree Judas used to hang himself. The other legend says that the tree blushes because of the betrayal of Christ by Judas. Regardless of the legends, we always expect our redbud to bloom for Easter.

Toyon

If one ventures into the foothills surrounding our valley, a definite change in the kinds of vegetation is noticed. The open grasslands with occasional interruptions by oak trees give way to more persistent plants of a type we call chaparral. The Spanish description of "chaps being needed to ride through it on horseback" gives you a good idea of the character of this plant zone.

There are many fascinating plants found in chaparral but none more interesting than toyon, *Heteromeles arbutifolia.* This handsome shrub is easily identified during the Christmas season because of its bright scarlet clusters of waxy berrylike fruit. In other seasons only the dark green oblong leaves with their rough edges give a clue to its name. The leaves are from 2 to 4 inches long and are a lighter green on the underside. In summer, clusters of small white flowers nestle among leaves of the evergreen branches.

This very attractive native plant is often grown as an ornamental in valley gardens. It is very popular with gardeners because it is not thorny like the common Pyracantha with which it can be confused.

Because of its popularity, toyon is known commonly by two other names—California holly and Christmas berry. It is this beautiful plant that gave the film capital the name Hollywood. Formerly, the hills surrounding the new community had a bountiful sprinkling of these shrubs. This obviously is not true today. Most people would say that the name Christmas berry was given because the berries ripen in the Christmas season. I like to think of the berries as a Christmas present from nature wrapped in lush green leaves to brighten up an otherwise drab span of vegetation.

Aside from its beauty the berries are eaten by birds. In the past the berries were also eaten by the Indians and Spanish Californians after various kinds of processing. Sometimes they were roasted and ground, sometimes boiled, and sometimes a cider was made. They were rarely eaten raw.

If you have room in your yard, try planting some of these fine shrubs. They can be started from slips or berries and will bring you years of enjoyment.

Poison Oak

Many people are curious about the ecological part poison oak plays, because of the tremendous problems poison oak causes with so many people, including me. It seems inconsistent that animals are not in the least bothered by contact with it. In fact, the leaves are actually eaten by bears, muskrats, rabbits, rats, mice, and mule deer.

The seeds are taken by many of our birds like the pheasant, Oregon junco, magpie, mockingbird, as well as many of our sparrows, thrushes, warblers and woodpeckers. It is interesting to note that flickers and several other woodpeckers are very fond of ants and can tolerate their formic acid. These same birds are also very partial to the berries of poison oak. Since the berries often persist into winter, they serve as an important wildlife food supply at a very critical time.

All of this is small solace to the poor suffering, susceptible individual that has come in contact with some part of the plant. If you are susceptible, itching and redness usually appear within 12 to 24 hours. Scratching in any stage may cause a secondary infection that will retard healing. Often in the later stages the infected area may "weep" or drip a pusylike fluid. This material will not give others poison oak or cause the irritation to spread. The uroshiol is absorbed to cause the irritation and cannot be passed along after absorption. One can get poison oak from clothes, boots, animals or anything else that has come in contact with the plant.

For treatment, my advice is to go to your doctor, but I can give you some ideas on how to prevent it.

My first advice is to learn to recognize it and avoid it. If contact is made, however, shower with a non-oily laundry soap. I take a cool soapy shower first, then follow with a separate soaping of warm water. In the field I like to carry rubbing alcohol and sponge off with it as soon as possible. Even though I am very susceptible I have had excellent results with these methods.

If you are forced to be around poison oak and are susceptible, see your doctor about immunization shots. They are usually quite effective. If you want to enjoy the out-of-doors learn about poison oak, and teach children to avoid it. As an ex-scoutmaster, I found I could teach my boys to avoid it during the day, but at night they mistakenly thought it wasn't there. Don't wander in poison oak country at night. You don't have a chance.

Poison Hemlock

One of my goals as a teacher and writer is to make people feel at ease in the out-of-doors. I try to dispel their fears of snakes and other misunderstood creatures whenever possible. It is necessary, however, to understand there are some dangers in nature.

One of these dangers is a rather common plant in our valley called poison hemlock, *Conium maculatum*. The stout stems of this plant may be four to eight feet high or even higher. The leaves are triangular in their overall shape, one to two feet long, and very dissected or fernlike. In fact, clumps of the leaves look like ferns and are sometimes planted in flower gardens for their lacy vegetation. This member of the carrot family has tiny white flowers in broad clusters. The oval seeds are grayish-brown with conspicuous wavy ridges. The root is a fleshy-white tap root that is usually unbranched. All of the above descriptions can fit many other plants, some poisonous and some not. The characteristics I use to tell the many wild carrotlike plants from poison hemlock are its purple spots on the stem and the mouselike odor of its vegetation. These are very distinctive and leave little room for doubt.

The deadly qualities of poison hemlock are indicated by its name. It was introduced from Europe and thrives in shady or moist ground in the lower elevations of California. It is also known as deadly hemlock, snakeweed, poison parsley, wade thistle, poison stinkweed, spotted hemlock, and California or Nebraska fern. All parts of the plant are poisonous although the toxicity of the parts may vary from time to time. The seeds could be mistaken for anise or dill but should not be.

This plant was used by the ancient Greeks as a means of quick death for disposing of prisoners. This is probably the hemlock Socrates drank. Water hemlock, another similar poisonous plant often found in the higher elevations, has many similar qualities. It is reported, however, that poison hemlock kills without convulsions while water hemlock produces convulsions. I can do without either.

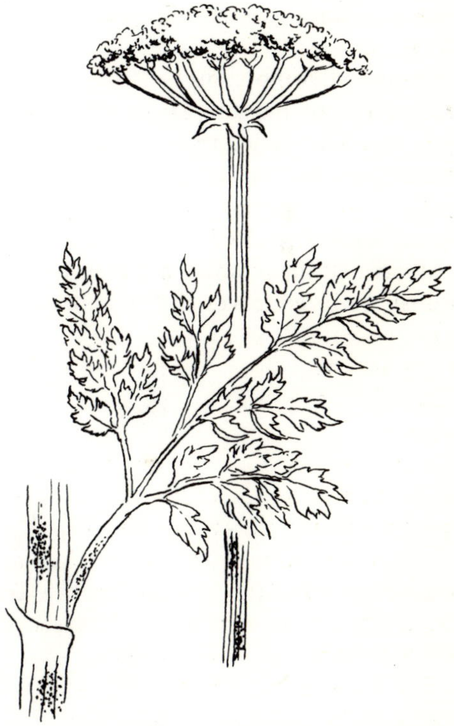

Poisonous Plants

It is always interesting to me to observe the way many people respond to the dangers of nature. Parents will often caution a youngster about a cliff, and yet be relatively unconcerned about automobiles and trucks that thunder along the sidewalks close to where their child is playing. Usually, medicines and poisons of the house are kept labeled and out of reach of children, but often poisonous plants are unknowingly planted in the yard where they are quite accessible to the curiosity of children's taste.

Two of the most poisonous plants often found in the garden are oleander, *Nerium oleander* and castor bean, *Ricinus communis.*

The attractive blossoms of oleander can be white, pink, or red and are borne in great profusion. This, along with the fact that it is an evergreen, makes it a very desirable shrub for the yard. It is also commonly used as a divider between lanes of our freeways.

All parts of the oleander are poisonous including the dried leaves. Even smoke from the burning plant affects some people. Poisoning has also resulted from using the branches as weiner sticks.

The poisonous effects of the plant are similar to those of the drug digitalis, and immediate treatment by a physician is necessary.

Castor bean is quite widely used as an ornamental and is even grown as a crop for the oil in the seeds. The large leaves, the unique shape of the plant, and the beautiful mottled seeds cause many people to plant it and not know what it is.

As few as three seeds can kill a child and six can kill an adult. The seeds must be broken or chewed, since the poison ricin is contained in the fleshy part of the seed. If they are swallowed whole, there is probably little danger in them. If you think a child has swallowed a seed, make him vomit if he is conscious, wrap him up and take him to a doctor or hospital. There is no antidote for the poison.

California Poppy

Many times when I try to express the deep feelings various subjects create in me I feel very insignificant. Such are the sentiments I have concerning the beautiful California poppy, *Escholtzia californica.*

The beauty that this flower provides Californians is beyond description. It seems to reflect all that we admire in our bountiful state and serves as a luscious frosting to an already providential cake.

The poppy family as a group is very showy with large conspicuous flowers. Their stems usually contain a milky or colored juice. They are common in temperate and subtropical areas, especially in eastern Asia and California.

The most noted poppy in America is our own golden California poppy. Its blazing beauty has endeared it not only to Californians but to people in many other parts of the world. It has been planted in gardens of Central Europe, India, and Australia where its dazzling displays are now naturalized.

In the United States it is found from lower California north to the Columbia River and east through Nevada, Utah, New Mexico and Arizona. It is in California, however, that the breathtaking panoramas have captivated all who behold them. The name of the Golden State was largely due to the countless rolling hills and valleys covered by glittering golden blooms. The Golden Gate got its name from the poppy-covered hills forming the entrance to San Francisco Bay. It is only fitting that this plant was selected as our state flower.

The California poppy can reach a height of two feet and have blooms two to four inches across. They may be from a flaming orange to golden yellow in color. In order to spread its petals it must push off a tight-fitting cap called a calyptra, and once the flowers have spread, they never close again. If picked, the flowers soon drop their petals and so are useless for decorations. The California poppy, as with most of nature's players, must be enjoyed where nature intended.

Johnny-Tucks

The leaves pinnately divided into filiform divisions; spikes slender, flat topped; bracts much shorter than the flowers; corolla eight to nine lines long and sulphur-yellow except the dark purple subulate galea, its filiform tube at least twice the length of the calyx.

The botonist's description you just read almost seems to be a foreign language. It is difficult to believe that what you read accurately describes a very common spring wild-flower called Johnny-tuck, butter-and-eggs or ham and eggs.

The Johnny-Tuck, *Orthocarpus erianthus,* is a member of the owl clover group in the Figwort family. They are four to six inches high with reddish, branching stems. The threadlike leaves are up to two inches long.

The yellow flowers are inflated and have dark purple edges and centers. The flowers are surrounded with what looks like purple whiskers.

Those beautiful bits of glittering gold start blooming in March and can still be found in May. They are so very common that many people do not bother to take a close look at them. The broad golden carpets that they create, however, can hardly be missed.

Johnny Tucks are most prevalent in the valleys and low hills of California up to 2,000 feet. They are rare south of San Luis Obispo and Kern Counties. Well-drained, open fields with a sparse grass cover seem to encourage their growth best.

If there is a vacant lot near, you should be able to find some of these gorgeous golden nuggets blooming there in early spring. They definitely play a part in making California a golden state.

Lupine

Although most observers through nature's window are busily engaged watching the actors, many of us are quite appreciative of the beautiful backdrops nature creates for her scenes. The open fields and roadsides are often the stage for dramas of nature that go unseen by human eyes. The settings are there for us to enjoy, however.

Many of the plants that compose the sets are beautiful in their own right and help make spring a thing of beauty. One of the showy expanses of flowers is the lazy, luscious lupine recognized by almost everyone.

Lupines can have either woody or non-woody stems. The leaves are on a long stalk and have three to seventeen leaflets arranged like fingers. The flowers are often arranged in a series of circular patterns up the tip of the stem. They are usually blue or purple—rarely white, yellow or red. In shape they are similar to the sweet pea. Lupine is a member of the pea or legume family.

The individual kinds of lupine are many and number nearly 200 species in this country. In Texas, they are called blue bonnets and are protected as their state flower. In California, we may have up to eighty species growing in habitats from arid plains and deserts to moist stream banks.

The seeds of lupines are valuable as wildlife food and are utilized by quail, Clarks nutcracker, mountain beaver, ground squirrels, deer, rats and many others. Although some species are poisonous, others are used as forage. The alkaloids, lupinine or spartenine, are found in the poisonous ones. Animals feeding on them will probably froth at the mouth, have trouble breathing, and go into convulsions that often end in death.

Understanding the lupines can teach us a valuable lesson of both ecology and life. The name lupine comes from the latin word *lupus* meaning "wolf" because it was thought the lupines robbed the soil of its fertility. This idea arose because the lupines tend to grow on poor soil. If planted in rich soil, most will not survive. As a legume, they enrich poor soil rather than depleting rich soil. Cause and effect relationships should be kept in mind, not only ecologically but as a way of life. This important concept is often ignored. Are you an offender?

Wild Mustard

The plentiful rains of spring create an aura of freshness and renewal in all of nature's world. When one opens the door in the morning and is greeted by the powerful rush of perfume from a gentle night rain, the feelings one has cannot be described. Spring is almost as if the plant world responds to this same refreshing stimuli and reflects its inspiration by covering the valley landscape with a colorful patchwork of showy flowers. Some of the flowers that nature uses are often of great singular beauty, but more often the dazzling displays of brilliant color are created by rather ordinary blooms. So it is with the wild mustards of the Great Valley. One of the first blazes of conspicuous color is the result of blooming fields of golden yellow mustard.

The mustards are native of Europe and Asia and so are introduced here. There are some eight kinds that have become established—some on purpose and some by accident. As a group they have rather large lower leaves that are variously lobed. The flowers are yellow with four spreading petals and may be from 1/4 to 1/2 inch long. The fruit is an elongated cylindric pod that has a beak at the tip. Both the foliage and pods have the characteristic pungent mustard taste and smell. The mature plants often are 4 feet or more in height. Although the mustards are widely spread in the United States, they are of importance only in the Pacific region and particularly in California.

Many of the mustards are well-known food plants. They belong to the group called Brassica which is the Latin name for cabbage. Cabbage, kale, cauliflower, broccoli, brussels sprouts, turnips and kohlrabi are all members of this group. Even our wild species are very tasty. They should be gathered while tender and boiled as greens. The flowers can be used also. If the flavor is too strong, pour off the fluid and boil for a few minutes in new water. Repeat if necessary.

White mustard seeds (not found here) are the source of the old mustard plaster. Black mustard seeds (common here) are the source of our table mustard.

Legend has it that the Mission Fathers spread the mustard seeds as they traveled from mission to mission. From this the seeds have come to mean good luck to those that carry them. Why don't you gather a handful and try it. Who knows?

Cheeseweed

In the wet season the rolling green landscape of our valley looks like a well-groomed lawn from a distance. When examined more closely, however, it is not grasses alone that create the lush green expanses. There are actually many plants that compete with the grasses, and one of these is known as cheeseweed, *Malva parviflora*.

Cheeseweed is a member of the Mallow family and was introduced here from Europe. It is very abundant in California and is common throughout the United States and southern Canada. It is a weed of waste places, gardens, and vineyards. It is sometimes used as a cover crop in orchards.

This coarse biennial grows 1 1/2 to 3 feet high and has roundish leaves from 1 to 5 inches across. Each leaf is composed of 6 to 7 shallow lobes. The small pinkish flowers are borne in dense clusters arising from the leaf axils or joints. The name cheeseweed comes from the fruit that is shaped like a little cheese. It is round and about one-half inch in diameter. It is less than one-fourth inch thick. When ripe, it breaks into about 11 pie-shaped pieces.

There are many other mallows somewhat similar in appearance. There is dwarf mallow, bull mallow, high mallow and several others. The distinctive characteristic of cheeseweed is the reddish spot on the base of the leaf. All of the mallows may be boiled as a green and eaten. Although I have not eaten them, I understand one should not eat large amounts as they can cause a digestive disorder.

If chickens eat the vegetation or seeds of cheeseweed, the whites of their eggs will be pink. (Probably the fatty acids malvalic and sterculic are responsible.) Evidently cheeseweed can cause a disease known as "shivers" or "staggers" in livestock. It can affect horses, sheep, and cattle and cause severe muscular trembling in them after strenuous exercise. Probably only the fresh plant is toxic.

Watch for this interesting plant. If the fruit is still green, peel it and taste it. You will find they have a pleasant taste.

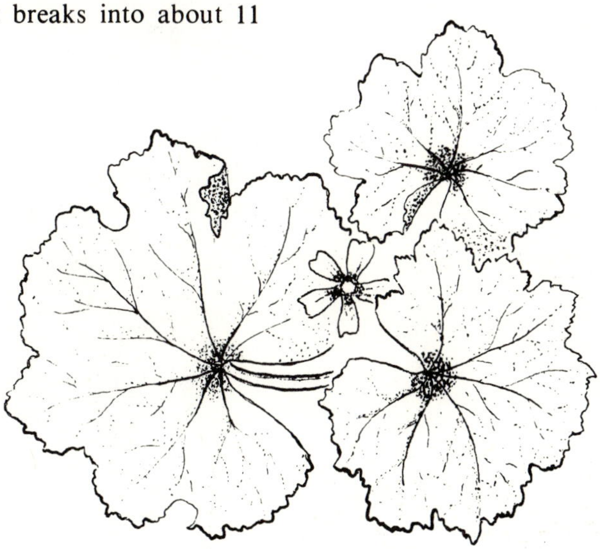

Chickweed

Early spring is an exciting time of year for everyone but especially for nature enthusiasts. The lush green blanket that dresses the earth seems to appear like magic after the warm spring rains. It is as if the glistening sun inspired Mother Nature to put on her finest robes.

If one looks at these robes closely, he will find they are composed of myriads of tiny green plants. The multitudes of individual species provide endless variety for those who study nature.

One of the most common plants in early spring is common chickweed, *Stellaria media.* This lush plant is very common while the weather is cool and damp. It disappears from the hot dry places very quickly. The one-inch leaves are borne on a droopy stem and are found in pairs. There are several stems in a loose diffuse mass that may rise to a height of 16 inches. There is an alternating straight, single line of hairs that runs up the stem. It is very conspicuous when examined with a magnifying glass.

The whitish flowers are 1/3 of an inch across with pale green sepals. They look like stars, and the name *stellaria* means "star" in Latin. The tiny petals are deeply divided which gives the flowers a very delicate appearance.

This introduced plant is very widely distributed throughout the world. It evidently followed European settlers wherever they went. We generally refer to it as a weed because it grows in places where we don't want it. We find it in our gardens, lawn, meadows, etc. It is one of the easier weeds to pull, however, since it is so tender and one handful can pull up a large clump.

Chickweed receives its common name because chickens and many birds eat it. The seeds are a favorite food of doves, quail, sparrows, Brewer blackbirds, and many others.

It is a very edible pot herb and salad green. The taste is very mild when cooked. For me, it needs some of the stronger greens like wild mustard, Shepherd's Purse or watercress to spice it up a bit. The chickweed should be gathered when young and tender, before flowering is best. The other greens should be cooked about 10 minutes, and then add the chickweed for 2 minutes more. Season with salt, butter, pepper and chopped onion. Chickweed can be used as a salad green by itself, but it is better when mixed with other greens.

Be careful where you gather this green but try eating them. I think you will be pleasantly surprised.

Chicory

Spring and early summer provide participants in nature's passing parade a constant variety of interesting and beautiful plants. Of the many found in weedy areas, none is more striking than chicory, *Cichorium intybus.*

Although it is now commonly thought of as a weed, it was and is often a garden plant. Introduced from Europe, this plant has several well-known uses that you may already be familiar with.

It grows as a perennial from a deep tap root. The erect stem is 1 to 3 feet high and has many dandelionlike leaves 3 to 8 inches long around the base. Smaller leaves also grow up the stem. The shape of all the leaves is somewhat variable.

The quite obvious feature of chicory is the attractive blue flower. These beautiful wheellike blossoms often make a blazing display when hundreds of plants are present. In fact, they are sometimes called blue sailors or blue weed. The flower or head is up to 1 1/2 inches across and has notched tips. They are rarely white or pink. Normally, these flowers stay open only till noon, but on cloudy days they may remain open longer.

Millions of pounds of the processed roots of chicory have been used by people as both an adulterant and substitute for coffee. Some prefer it to coffee, while others have used it to stretch a coffee supply. This was especially true during the Civil and World Wars.

To make your own chicory supply, dig and scrub some roots; roast them slowly in a partially open oven until they crumble between your fingers. Then grind and store in a closed container until brewed as coffee. Use lesser amounts than coffee. Chicory can also be obtained at most health food stores.

The leaves can be gathered while the plant is quite young and are similar in many ways to dandelion leaves. The young leaves are pleasantly bitter but become very strong as they get older. Gather them well before the plant flowers and cook as pot herbs. Because of the edible qualities, chicory is also called succory or chicory lettuce.

The milky sap of the plant tastes bitter and if milk cows eat much of it, their milk tastes bitter. Because of this, farmers do not like it in their pastures. You just can't please everybody.

Dandelion

Several of the plants that we find around our house are missed because they blend into a greenery background. Such is not the case of the quite conspicuous dandelion, *Taraxacum officinale.* This very common weed is so well-known it is hardly necessary to describe it.

The golden yellow flowers tell us when we need to mow our lawn, and the leafy rosettes indicate that we had better get our lawn in better condition. It is difficult to get the long tap root out of a lawn without actually digging it up although there are very effective poisons that can be injected into the root to kill it.

The rather tender leaves are right on the ground and may grow to ten inches in length. The bright one to two inch flower is borne on a hollow stalk that may be very short to more than a foot high. When the flower stalk is broken it exudes a rubberylike milky substance. The flower ripens quickly into a beautiful fuzzy head. Each seed has its own private silky sail that opens to ride the wind. As a boy, I remember blowing the seeds from the head and with each puff saying, "She loves me, She loves me not."

This plant was called dandelion because of the similarity in shape between the points on the leaf and the teeth of a lion. In France, the name *dent de lion,* which means "lion's tooth," was given to this plant. To the English, it sounded like dandelion, and that is what we call it.

The dandelion is an introduced plant from Europe, but it is well established here. It is sometimes cultivated for a drug found in the root. The root can also be roasted and used as a substitute for coffee. The roots have a medicinal effect and have been used for centuries throughout the world as a tonic, mild laxative, and diuretic. An excellent wine can be made from the flowers.

The leaves may be eaten raw in a salad if they are not too old and bitter. They can also be cooked. If there is a bitter taste, change the boiling water two or more times.

If you have never tried any wild foods, try the dandelion. It is excellent in stew, with ham or bacon. I am sure you will enjoy the change.

Red-stemmed Filaree

A walk in the dry open fields of our valley can be a very unpleasant experience if you are not prepared for it. High boots and heavy trousers are a necessity to keep so-called stickers from sticking us. One of these "stickers" is from a plant called red-stemmed filaree, *Erodium circutarium.*

There are eight species of filaree found in California, and only two of them are native. These members of the Geranium family all have a long beak on their fruit. The name erodium refers to this characteristic and is from the Greek word *erodios* meaning "a heron." From this comes an often-used name storksbill.

Red-stemmed filaree was introduced very early from the Mediterranean region. It is either an annual or biennial that can have prostrate or erect stems 3 to 12 inches long. In the early stages, the leaves form an often reddish rosette close to the ground. The leaves are very frilly and somewhat fernlike. As they get older, they turn dark green and are one-half to four inches long. The pinkish flowers are similar to other geraniums although smaller. The flower is almost one-half inch across.

The long beak on the fruit is familiar to almost everyone. Children often punch one green fruit through another and make a scissorslike toy. Because of this, it is sometimes called scissors plant. Another interesting thing to try with the fruit is to find one almost ripe and pull the covering from the base. Then carefully strip out one seed and its beak. Tape the seed down on the edge of a card and watch it wind up like a spring. In this shape it looks like the stickers that annoy us so often. The twist along with hairs on the seed serve as a self-planting device. The humidity causes the spring to open and close causing the seed to drill into the ground and plant itself.

Red-stemmed filaree is the most common filaree in our valley, and in early spring is an excellent forage plant for livestock. Very young plants can be cooked as pot herbs or used in salads.

In 1844 John Fremont wrote in his report of the valley that filaree "covered the ground like a sward" and "instead of grass, the whole surface of the country closely covered with it." When Fremont crossed the Sierra, filaree growing in the foothills gave his animals the strength to reach Sutter's Fort. So much history for such a small plant.

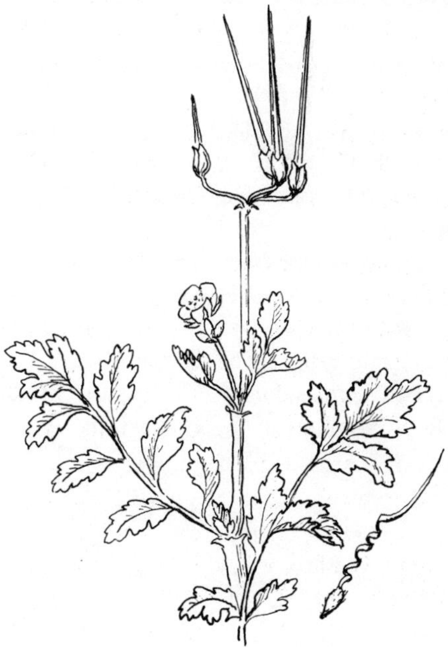

Horehound

In spring as you poke around weedy waste areas, there is a whitish plant with soft woolly leaves that is often quite prevalent. I am referring to horehound *Marrubium vulgare.* Horehound has erect stems that may grow to 30 inches. The woolly white stems are squarish as are other members of the mint family. The leaves are one to two inches long, oval shaped with scalloped edges. They are greenish on the top but densely woolly underneath. The leaves grow opposite one another up the stem. The white flowers are grouped in very dense clusters above the leaves.

Horehound is not a native species but has been introduced from Europe. It is now widespread throughout the United States but is most abundant on the Pacific Coast. It is one of the most common weeds in California and has been cultivated in the United States in the past. A good crop would yield about one ton of dry herb per acre—worth about 7 cents a pound. Although it is not recognized as an official drug, about 100,000 pounds are imported into the United States each year.

The name *marrubium* as part of the scientific name is from the Hebrew meaning "bitter." Both the leaves and stems of this plant are extremely bitter. Since we always think of medicines as being bad tasting, this is probably the reason there are so many medicines and home remedies that have horehound in them.

Remedies were made to treat colds, rheumatism, dyspepsia, and high blood pressure. A strong infusion was used to treat worms. Perhaps the greatest use was to sooth sore throats and coughs by putting an extract in candy. Horehound candy can till be found.

If you would like to taste this interesting and historical plant, dry it and make it into a tea. It will be good, nutritious, and bitter. In large doses it becomes a laxative. Perhaps for you it might relieve another ailment—boredom. Try it when you get a chance. You might like it.

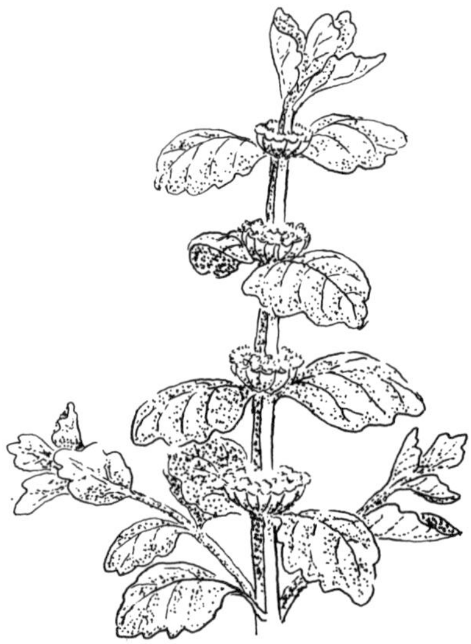

Mugwort

Anyone interested in the out-of-doors in our valley will sooner or later come into contact with a very common plant called California mugwort, *Artemesia douglasiana* formerly *A. vulgaris.* This plant is widely distributed throughout the state and is generally found along streams, rivers, irrigation ditches, highways, and other moist land.

It may grow to six feet although it is generally around three feet high. The stems generally grow up straight from a running rootstock. The plant is a perennial. The leaves are somewhat oblong, lance shaped, or oval. The lower leaves are usually cleft one or two times along each edge while the upper ones are not. In summer, the cleft leaves often give way to small uncleft ones. The top sides of the leaves may be a lush green to a grayish color and the underside are generally silvery and cottony below.

The flowers are rather difficult to see because they are so small. The dense spikes of the flowers are erect. There is a papery greenish sheath or involucre that encloses 20 to 30 of the tiny flowers.

This close relative of sagebrush has a rather pleasant, pungent odor all its own. You may smell the plant even before you see it. If you suspect a plant of being a mugwort, crush one of the leaves and you will know for sure. The smell is strong and characteristic.

As often seems the case, Indians and pioneers alike used plants with strong smells or flavors for medicinal purposes. This was true of mugwort. The juice of the plant was used to treat or soothe skin disorders. It was also common practice to place a crushed piece of leaf in each nostril to help headaches and colds. Indians used it much the same as we use an olive branch, as a symbol of peace. Watch for this very characteristic plant. You won't have any trouble identifying it.

Miner's Lettuce

I have found that almost everyone is interested in plants in one way or another. Some are interested in beautiful flowers, others in odd plants like cactus, some in those that are fragrant and others for use in landscaping. It seems, however, that everyone is interested in one thing—whether or not you can eat them.

We take for granted the fruits and vegetables we eat but are fascinated when we find plants growing wild that are edible. Although many plants can be prepared in special ways and made edible, some are ready to eat in the field. One of these that is common and very tasty is miner's lettuce, *Montia perfoliata.*

This lush succulent plant is a native of the west coast and is a common weed along waterways and in orchards, gardens, and vineyards. It is a smooth low annual plant from 4 to 12 inches high and very easy to identify. If you look for the disclike top leaf that has the stem passing right through the middle, you can't make a mistake. The flowers are found above the leaf on a short extension of the stem. They are whitish to pinkish in color and are also edible. The leaves that arise from the base are long, narrow, and broaden at the top.

The early pioneers and miners made good use of this plant as a green. They could use it in salad without preparation or cook it as spinach greens. As far as my own taste goes, it is at least as good as lettuce and more fun to eat.

If you would like to try this plant, late winter and early spring is a good time. At this time of year the plants are very succulent and tender. They should be gathered from a place where there has been no spraying or dogs, and then washed well. Eat it just that way or with a little salt, vinegar, or your favorite salad dressing. It really is great, try it.

Purslane

Summer time and the high food prices of today make it possible and worthwhile to explore other sources of food. Many people now garden seriously to help defray the high costs of living. While gardening, many valuable food plants are often pulled out as weeds. One of these is the well-known weed common purslane, *Portulaca oleracea.* It is also known as wild portulaca, pussley or pursley. It is thought to be a native of India that was carried to southeast Asia and then to Europe. From Europe it was introduced into the New World where it is now well established. It is widely distributed throughout the West in both cultivated and waste ground.

Common purslane is a semi-succulent fleshy annual that is quite smooth and forms a matlike growth close to the ground. The stems are often reddish and the numerous leaves are alternate, quite thick, and somewhat wedge-shaped with rounded tips. The small yellow flowers open only in the sunshine. The numerous seeds are contained in a globular capsule that breaks open around the middle. Birds and rodents that eat the seeds help spread the plant prolifically.

Almost every garden has some of these quite edible herbs, and you are missing something if you do not try them. The young leaves and stems are best, but the entire plant except the roots can be cooked and treated like spinach. The young leaves and stems should be washed well, and then can be cooked or eaten as a salad.

Purslane has one of the highest iron contents of any green vegetable. That is a pretty good record for a plant that is 92.5 percent water.

Shepherd's Purse

There are many fascinating plants that many of us never really notice. We pull them from our gardens, walk on them in the fields and even mow them in our yards, but we may never really see them or know what they are. Such a plant is Shepherd's purse, *Capsella bursa-pastoris.*

This plant was introduced from Europe and has become naturalized throughout the United States. It is found in gardens, along roadsides, in waste places where there is adequate sun and not too much water. It also does well in places where the soil has been packed. I have not found it growing above 7,000 feet in our Sierra Nevada.

The flowers of shepherd's purse are small and inconspicuous but may appear throughout the year. The fruits, however, are more obvious and are taken to be leaves by many. Although there is great variety in the shape of the tiny fruits, the predominate shape is heartlike. They are attached to the rest of the plant by their point or apex and are about one-third of an inch long. The plant may grow to 20 inches with a cluster of quite variable leaves at the base. These are often deeply cut as shown by the sketch. The whole plant is very hardy and can withstand severe drying and chilling.

The tap root often penetrates the soil quite deeply and is quite woody. The plant is generally killed, however, when the tap root is cut by a hoe.

The seeds are eaten by several birds like horned larks, goldfinches and quail. Chickens and other animals may eat the whole plant. When cooked as a pot herb, it is inferior to many other greens but has a cabbage taste. The Indians roasted the seeds for a nutritious meal, and an infusion of the leaves was used for bruises and taken internally to stop bleeding.

The name shepherd's purse refers to the similarity of a purse carried at the waist. The term capsella means little box, while bursa means sack and pastoris refers to shepherds. Evidently the fruits have impressed people for a long time. Why not have a close look for yourself?

Salsify

It is fascinating to watch the changes that spring brings about in the rather drab dress of winter. To the observer through nature's window, weedy waste areas turn into veritable flower gardens for those who appreciate them. One of the most intriguing of the plants that may appear is salsify *Tragopogon porrifolius.* June through August is the normal time for flowering.

The large purple flower usually stands well above the other plants. It is borne on a generally unbranched stem two to four feet high. The numerous smooth, narrow, grass-like leaves are eight to twelve inches long. Their bases firmly clasp the stem. The flower is two to four inches across when open. They open early in the morning but close into a small green case by noon. The stem is usually swollen and hollow just below the flower.

The flower turns into a beautiful giant dandelion-like head of seeds. They are light-brown and quite striking. You will have no trouble recognizing them. The seeds are easy to gather. As the seeds form, they have long white hairs attached that form the sails. These hairs gave rise to the Greek name *Tragopogon* which translated means "goat's beard"—another name.

This rather widely distributed biennial or perennial plant is an introduction from southern Europe and is now found in various parts of the world like Asia and North Africa. Cultivation has been known since the year 1600 but the plant is not commonly grown now. The root is known for its unique oysterlike flavor and somewhat resembles parsnips. Salsify is often called oyster plant.

To grow, they are planted in rows 12 to 15 inches apart and thinned to two inches between plants. They require a long season, and the roots are normally dug in winter. They are stored like beets, turnips, or parsnips and are marketed in bunches of about ten.

The roots may be eaten raw or cooked. The stems and leaves may be eaten as pot herbs when only a few inches high. The milky juice from the stems was considered a remedy for indigestion, and the distilled juice was applied with pads to bleeding sores and wounds. There were several other uses.

Watch for the plant in summer and put a stake by it so you can find the root later. Try it and see if you like it.

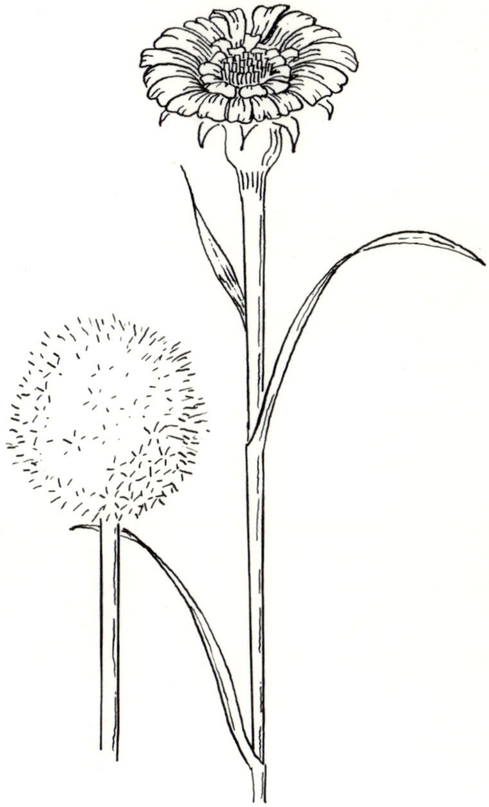

Soap Plant

Soap plant, *Chlorogalum pomeridianum,* is a plant that attracts attention in different ways in different seasons. In spring, the rosette of leaves on the ground is very obvious even though it is surrounded by the lush spring growth of grasses. The individual basal leaves are up to a foot long and about one-inch wide. Their edges are generally wavey.

The leaves sprout each spring from a very interesting bulb. It is three to four inches long and thickly covered with coarse dark fibers. It is bottle-shaped and deeply set in the ground.

The stem of the plant rises from the center of the leaves and when young, it reminds me of asparagus. It grows from two to ten feet tall and has a very loose arrangement of flowers called a panicle. The flowers have white petals that are quite narrow and about three-fourths of an inch long. As the petals get older, a purple vein appears. Many people ask me if this plant ever blooms. The reason for the doubt is because the flowers open in late evening. When open, the petals curl back over the capsule. The flowers appear in early summer, and the tall stalks are quite obvious growing above the grasses.

The leaves by this time, however, have shriveled and all but disappeared. Unless you followed the plant throughout the season, you might not connect the leaves with the flowers.

The name soap plant comes from the ability to use the plant for soap. If the fibers are stripped off the bulb and the heart crushed, an effective lather and soap can be produced. It is said to be excellent for washing hair.

The green plant and root can be crushed and used to stupify fish. It takes quite a quantity to work and is quite illegal. The bulb may be eaten if roasted and then peeled, or peeled and then boiled. The small young shoots should be slowly baked or steamed for a nourishing green. The leaves when young and fresh, may be eaten raw. This plant is also called amole.

This highly utilitarian plant is more common than you might think. It might be growing in the field near you. Take a walk some evening and see if you can find one in bloom.

Tomato

I am sure many people of the Sacramento area are as eager about their gardens as I am. It is always a difficult wait from the time we plant our seeds or plants until we harvest our first reward. Although I have planted several things in my garden, it is the tomato that I watch and wait for most.

I have found that almost everyone is familiar with the tomato, but few people are aware of its interesting history. The wild ancestors of the garden varieties of tomato are two tender perennials that are natives of western South America, *Lycopersicon esculentum* and *Lycopersicon pimpinellifolium.* From South America the tomato was introduced into the gardens of Europe as an ornamental. Its value as a food was not appreciated. This was probably because of its close relationship to many plants that are quite poisonous in the nightshade family in which the tomato is now classified.

Tomatoes were grown by Thomas Jefferson in Virginia in 1781. They evidently were grown as an ornamental and curiosity, however, because they were not known as an edible vegetable in America until after 1834. It was another ten years before it gained any popularity.

In these early times the popular names for the tomato were love apple and gold apple.

There are some poisonous aspects of the tomato of which gardeners should be aware. The vines and suckers have proved toxic to livestock. Cattle and pigs have been poisoned when garden cleanings were thrown into pastures so care should be taken with clippings.

Mistletoe

In our celebration of Christmas, the many traditions provide us with the basis of our holiday festivities. None of the traditions provide any more fun than a kiss under the mistletoe however.

There are many kinds of mistletoe, and they are all parasitic on a variety of trees. The species used in America is American mistletoe, *Phoradendron flavenscens.* It is generally associated with oaks, but it may grow on other deciduous trees.

The plant is shrubby, rubbery, and greenish barked. The leaves are pale green, rather thick and smooth, and easily broken. The white berries are eaten by birds like cedar waxwings, phainopeplas and bluebirds. Some of the seeds stick to the outside of the bird's bill, and when rubbed off spread the plant. The berries are reported poisonous to humans.

The parasitic needs of the plant are support, water and minerals. The first part of the scientific name, *Phoradendron,* is from the Greek with *phor* meaning "thief" and *dendron* meaning "tree". Even the word mistletoe is a corruption of a Saxon word *michtan* that means "different twig".

The early Cets used mistletoe as a charm against evil spells. Early Europeans used it in all kinds of ceremonies which led to its use during Christmas holidays. It is said to bring happiness, safety, and good fortune as long as it does not touch the ground. Perhaps this is why it is always hung overhead.

The kissing custom may be from an ancient legend that involved a goddess and her son. The goddess went to everything on earth and exacted a promise they would not ever harm her son. She contacted everything, that is, but the young mistletoe. Later, a trouble maker found out about the mistletoe and manipulated a blind man to throw a mistletoe spear at her son. Her son was struck in the back and promptly fell down dead, and the god of the underworld claimed him. His mother evoked so much sympathy the underworld god gave up her son. She was so elated that she gave out many gifts to the earth and even kissed the mistletoe in forgiveness.

Common Groundsel

Winter rains and warm sunny days stimulate a very lush growth of spring plants. The wild grasses and other plants carpet our landscape with green. In fact, many normally well-managed flower beds in springtime have an abundance of so-called "weeds." The term weed, however, is a word that refers to a plant growing in a place you do not want it to grow. You plant grass in your lawn, but when it grows in the flower beds it is a weed. I guess we are just hard to please.

One of the obvious weeds in many places, especially along roadsides, is a small plant that goes by various names. Its scientific name is *Senecio vulgaris* which means common groundsel. Other names it goes by are old man of spring, grimsel, simson and ragwort.

Since it was introduced from Europe, it has spread from Newfoundland to Georgia and west to the Pacific Coast. It is very common in California.

Almost everyone has noticed this 6 to 15-inch, well-branched plant. The all-yellow flowers are borne more or less erect. There are usually several in a cluster. The yellow heads have no ray flowers. The leaves on this annual are alternate and up to 6 inches long with jagged margins. When seeds are produced, they are in little white balls of fuzz.

There are at least a thousand of the groundsels or senecios throughout the world. There are about 38 species found in California. A characteristic useful in identifying the common groundsel is the cuplike structure that holds the flowers. This cup consists of 20 black-tipped bracts. Some of the senecios contain poisonous alkaloids, and common groundsel has been mentioned as, perhaps, poisonous to some animals. I can find no documentation of this in North America.

Take a look in your garden and see if you have any of these yellow-flowered plants. Look at the bracts and maybe you can now call this weed by name. You may be surprised at how many there are.

Mayweed

Mayweed, *Anthema cotula*, or dog fennel as it is often called, is somewhat similar to another plant called pineapple weed, *Matricaria suaveolens*. Both of them have very lacey, fernlike leaves. Pineapple weed generally appears earlier in the spring and the oval-shaped yellow heads are without the white petals present on mayweed. Because of the similarity of the leaves, many people assume the mayweed is just the pineapple weed in bloom. Such is not the case however.

Pineapple weed has a very pleasant smell and taste, while the smell of mayweed is quite unpleasant and the taste very bitter.

Mayweed grows to two feet high with a well-branched stem. The leaves are up to two inches long and about one inch wide. The head or flower is about one inch across with a yellow center and a ring of daisylike white petals around the outside.

This European species was introduced in the days of early Spanish settlement. It is widespread throughout the state and is usually found in wastelands, pastures, and along roadsides.

The herbage smells bad and its juice can evidently cause a burning irritation to wet skin. If milk cattle eat it, the milk will taste bad, and if eaten extensively, it can cause blisters on their skin.

Although often called chamomile, it is not the one raised as a drug plant. That one is garden chamomile, *Anthemis mobilis. Anthemis* is the Greek name for "chamomile." Mayweed is sometimes called fetid chamomile but has also been called white stinkweed, stinking daisy and the Spanish name manzanillo.

Although called mayweed, it generally starts blooming in late May and continues through August. It provides seeds for birds like goldfinches, juncos, sparrows and even the brown towhee. Everything has some redeeming feature. Watch for this interesting though smelly plant.

Wild Morning Glory

One of the very common weeds we find in our gardens is wild morning glory, *Convolvulus arvensis.* This plant is quite obnoxious to the gardener, but it does have some beauty in its own right. The snow-white blossoms with their delicate pink markings that open in the morning may look like snow on the ground when the weed becomes dense. It is then we can see why it is called morning glory.

This trailing perennial is widely distributed throughout the lower elevations of California. The somewhat triangular leaves are borne on slender stalks about one-inch long. When the plant grows under ideal conditions the blade of the leaf is broad but under adverse conditions the leaf blade is narrow. The funnel-shaped flower is white or pale pink inside, but when closed looks purplish.

This introduced European plant is often mentioned as the most troublesome orchard and garden weed yet naturalized in California. It is especially troublesome in rich alluvial soils. This gives rise to two of its common names—bindweed and orchard morning glory. The ability of this plant to climb and completely cover other plants enables it to choke out most competition.

One of the characteristics that makes control of this plant difficult is the delayed seed germination. The seeds may lie dormant in the soil for many years. This means that new plants can spring up even if there are no mature plants present. Another problem in controlling wild morning glory is the ability of any section of the root to send up shoots. I have been pulling one morning glory in my rose garden for years. It still comes up. If a field with morning glory is plowed, and its very deep tap root cut up in several pieces, each of the pieces can give rise to a new plant. The pieces of tap roots can send out very diffuse and long laterals. These rhizomes may form large masses of roots; hence, the name bindweed.

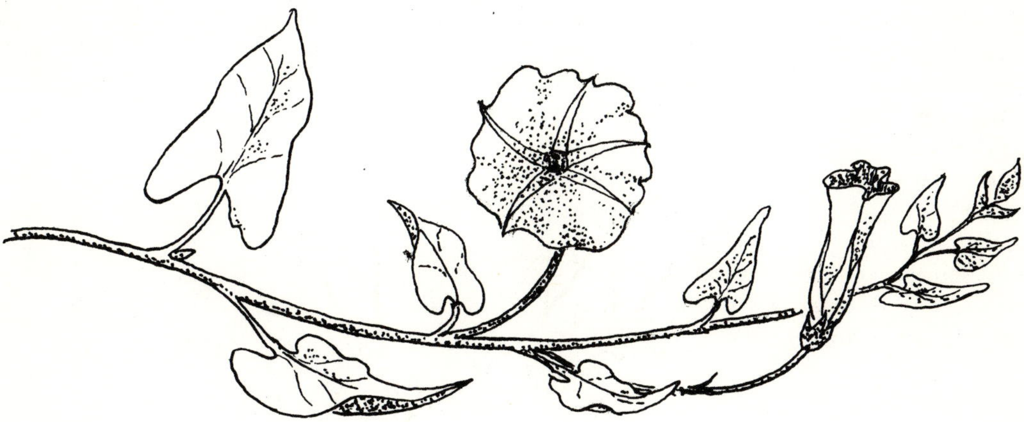

Common Mullein

Striking scenery can be seen in some of the most peculiar places in California. If you just look around, there are some fascinating views in very unexpected places. Normally one thinks of the ocean, mountains, forests, desert, or skies when one thinks of outstanding scenic beauty. We seldom think of dry stream beds or roadsides as possessing unique scenic potential, but it is there. It is there in a very peculiar plant called common mullein, *Verbascum thapsus.*

This plant is often more than six feet tall, and in spring and summer feels soft and woolly. There is a rosette of leaves at the base that are up to one foot long but get smaller as they appear up the stem. Often, the flower stalk rises three feet above the last leaf. It takes two years for the yellow flowers to appear. They are over one inch across and are densely crowded on the spike. The flowers open below and as the season progresses, they bloom on up the stem.

It is in the fall and winter, however, that this plant impresses me. This is the plant one sees growing along rocky roadsides usually in full sun. It has such a variety of grotesque shapes that it is difficult to believe they are all the same plant. The tips may fan and curl. There may be side stems like arms, and I have even seen loops. They are a photographer's delight with their countless fascinating forms.

It is sometimes called velvet plant and was introduced into America from Europe and Asia. It is no relation to the turkey-mullein that provides the seeds for mourning doves. The hairs are collected by hummingbirds for their nests, and the abundant seeds are utilized by goldfinches. Early settlers dipped the woolly leaves in grease and used them for wicks. The hairs may cause a skin irritation on some people. There have been many healing powers attributed to this plant in early days. It was used for cramps, convulsions, coughs, and toothaches just to mention a few. None have been verified as far as I know.

Watch for this intriguing plant as you travel. Perhaps as a friend, it will brighten your journey.

Turkey-Mullein

One of the great pleasures observers through nature's window can experience is the beauty some of the more inconspicuous partners in our environment possess. When one first starts observing our fellow earth inhabitants, he usually sees only the more obvious ones.

There is a plant that blends in so well along our roadsides that most people never distinguish it at all. I am referring to the very common turkey-mullein *Eremocarpus setigerus,* of the Spurge family. It is not at all closely related to common mullein.

Turkey-mullein is essentially a California plant that does not extend very far out of our state. It thrives in dry open areas like stubble fields, fallow (unused) land and roadsides. It is very prevalent in our valley and is most often seen by motorists along weedy roadsides.

This native plant is a low annual that may be 1 or 2 feet across and up to 1 foot in height. It is often mound-shaped and very attractive when looked at closely. The stems and leaves are covered with a dense coating of forked, bristly, star-shaped hairs that make the plant harsh and gray. The leaves are thick and three-veined, and from 1/3 to 1 1/2 inches long. They are oval to round in shape. The flowers are very inconspicuous.

The seeds of turkey-mullein are very valuable to wildlife. They are eaten by doves, quail, finches, sparrows, rabbits, ground squirrels and others. There is a narcotic poison in the stems and leaves, but it is not believed to poison animals that eat them. The hairs on the plants, however, have been known to cause "hair balls" and death in sheep and hogs. Indians used broken stems and leaves to stupefy fish. It is thought the hairs affect the gills so the fish can be caught by hand.

In late summer it is almost possible to identify turkey-mullein by the mourning doves that fly up from them as your car frightens them. Turkey-mullein provides a major part of the mourning dove's larder. That is why it is sometimes called dove weed.

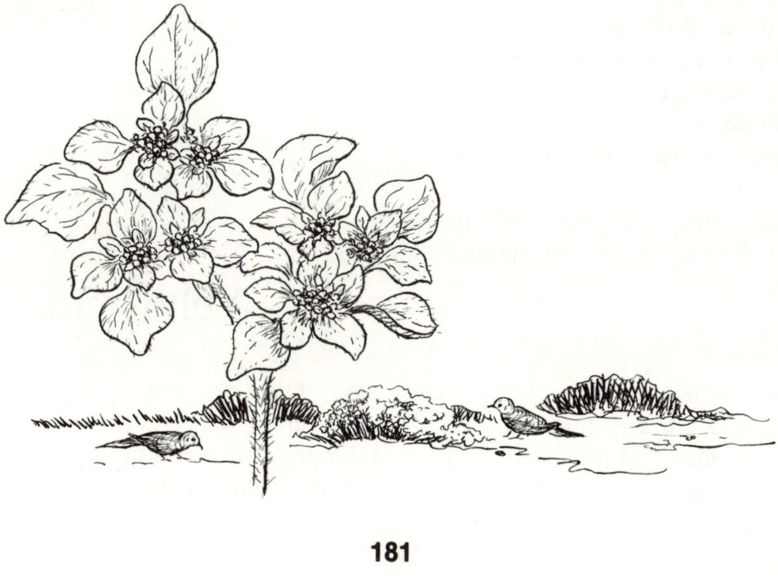

Puncture Vine

Bee stings, poison oak and sunburn are not the only problems of summer. A plant called puncture vine, *Tribulus terrestris,* must also be considered a real nuisance.

It was accidently introduced from the Mediterranean countries and has become quite widepsread in California. It is an annual that springs from a single tap root, and the bright green leaves and leaflets are arranged opposite each other. The stems radiate out from the root crown and branch freely. Small rootlets branch out from the tap root and form such an effective root system that moisture is garnered from soil that will support few other plants.

The plant is quite prostrate when it grows in open ground, but it may trail like a vine along fences or among other upright weeds. The stems are usually reddish and, like the small oval leaflets, are densely covered with silky hairs that give the herbage a silvery appearance.

The small bright yellow flowers arise from the same point that the leaves arise. The petals usually stay open only till noon or perhaps later if the weather is cloudy.

The fruit is a cluster of five spiny nutlets or burs. When mature, the nutlets separate and their spines serve them well as a means of dispersal. No matter how the bur or nutlet lies, one of the two spines points up ready to puncture feet or tires and be carried to another place. Even the toughest of bare feet yield to the one-fifth of an inch spines. Many bicycle tires are also punctured.

Many physical injuries are caused to domestic animals and people. I have even had the spines go through the sole of a shoe.

Watch for this plant in our valley. Take a look at the nutlet, and you will have no doubts why it is called puncture vine.

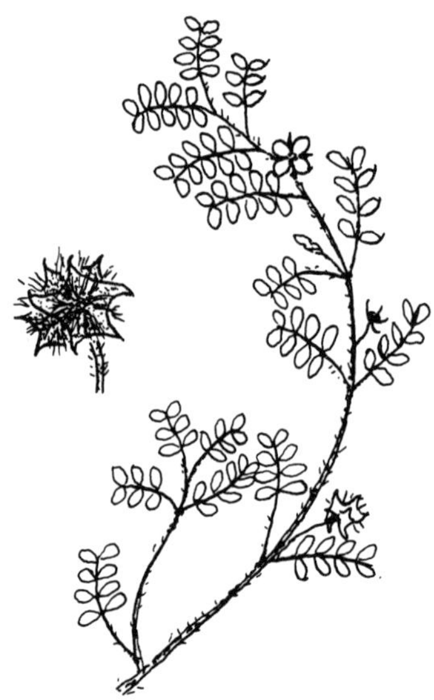

Yellow Star Thistle

The warm brown fields of our valley often look very inviting from a distance. To the uninitiated they would seem to be very pleasant to walk through, but such is not the case. There are many different plants that make up these dry fields, and most of them are harsh and scratchy. They are composed mainly of grasses, but there are other plants that are very conspicuous. One of these is the yellow star thistle, *Centaurea soltitialis.*

The plant has rigid spreading stems, one or two feet tall, that have thin winglike projections running along opposite sides. Most of the plant is covered by loose, woolly hairs giving the foliage a whitish cast. The leaves are long and narrow and feel soft and limber to the touch. It is the flower, itself, that is sharp and scratchy.

The flower heads are solitary and found at the end of the branches. They are bright yellow and have long, rigid, two and three-pronged spines that are quite sharp. During winter after the seeds have formed and dispersed, the spines fall and the stem dries up but generally remains standing. A gray puff of cottony material is all that remains of the flower. This characteristic dried weed is seen by almost everyone, but few realize it is what's left of the yellow star thistle.

This weed was naturalized from Europe and is sometimes called Barnaby's thistle. It has spread steadily since 1885 and is not eaten by livestock. It is very undesirable in cultivated areas, and its introduction and spread was generally as a seed impurity. Although it is a nuisance to most of us, it is well used by wildlife. The seeds are favorites of goldfinches, mourning doves, California quail, sparrows and other birds. Bees use it to make a fine honey. Watch for this common plant. If you don't, you may say "ouch!"

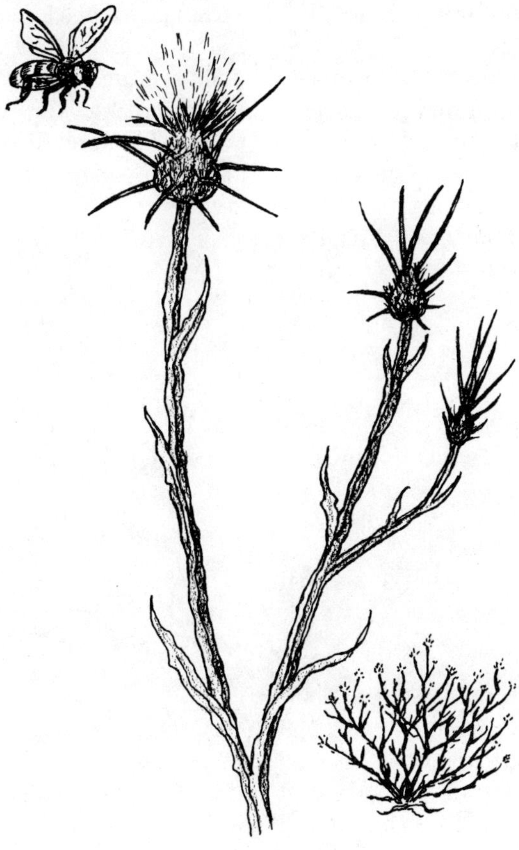

Cocklebur

It is intriguing to me that certain things in nature carry my thoughts back to the days of my childhood.

One nostalgic reminder is the cocklebur, *Xanthium strumarium* variety *canadense,* that is found in our valley. I am sure that you may wonder how this coarse plant with such spiney burs can fill one with nostalgia.

As a child, I spent many hours in Ohio gathering the burs of a similar but different species of cocklebur and fashioning them into baskets, statues, balls, stars, etc.

Our cocklebur is also known as sea burdock, sheep bur, bur-weed, clatbur and ditch bur. It was introduced from the Old World tropics and is now well established in all temperate and warmer zones.

It is exceedingly abundant in California along unmaintained irrigation and drainage ditches and on low marshy lands and pastures. It is common along Arcade Creek on the American River College campus.

Cocklebur is a coarse, rough, annual plant that is usually between two to four feet tall. The stems are somewhat angled, tinged with reddish brown or irregularly spotted.

The leaves are up to four inches long with the base somewhat heart-shaped. The rest of the leaf is variable in shape. It can be unlobed or have three or five lobes.

The edges of the leaf are coarsely toothed while the surfaces are rough and covered with minute bristles. They are dark green above and lighter below with three conspicuous nerves.

The burs are oblong or elliptic in shape and up to one inch in length. They are covered with hooked bristles. The burs have a pair of stout hooked beaks at the tip and each has two black seeds.

The seeds and plants are poisonous. The plant is more poisonous in the seedling stage.

The burs are an excellent example of adaptations for seed dispersal and hook on to almost any mammal. I know, I have carried my share.

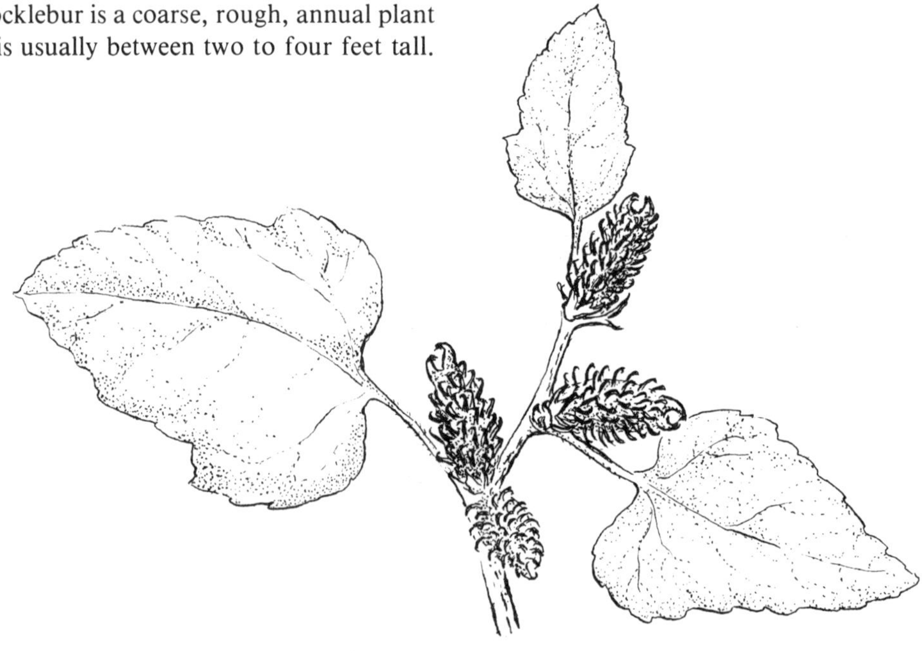

Index

As we become more and more involved in the ecological intricacies, we may see only the trees and not the forest, or see the waves and not the ocean.

For many of us it is the "forest" and the "ocean" that lift our spirit and give us the strength to face the more mundane aspects of living.

G

H

I

J

K

L

M

N-O

P

Q

R

S

T